BENEATH THE GLASSWATER SPRING

Written by Diane Kann

VOLANS

Brought to you by Volans Galaxy Press

Published by Kannceptual Creations LLC

An imprint of Volans Galaxy Press

ISBN: 979-8-951552-03-7

Printed in the United States of America

First Edition, 2026

Note: This work was originally published under the pen name DM Volans, which is a pen name of Diane Kann.

CONTENTS

DEDICATION

For every young heart that refuses to look away—
for those who listen closely when something feels wrong,
who question what others accept,
and who dare to uncover the truth, even when it runs deep.

May you have the courage of Avery—
to stand firm when the world feels uncertain,
to protect what cannot speak for itself,
and to follow the quiet voice inside that says, *this matters.*

This story is for you—
the ones who notice, who care, and who act.

Because sometimes, the greatest change begins
with someone brave enough to ask why.

Chapter One

The Morning Tide

The water didn't just feel lower; it felt tired.

Avery pulled her paddle from the spring's glass-clear surface, the blade slicing through the water with a sound like a wet slap rather than the usual, crisp parting of current. She drifted, the kayak nudging softly against a cluster of pickerelweed. Beneath her, the limestone vent—the "Heart of Glasswater"—usually pulsed with a vigorous, rhythmic surge that pushed back against the encroaching river water. This morning, the pulse was a tremor.

She looked down, her reflection staring back, distorted by the slight ripple of an underwater eelgrass bed. In her photography days, she'd learned to read the light, but in these past few weeks, she'd learned to read the rhythm. The eelgrass, usually combed flat and emerald-green by the steady, unrelenting force of the aquifer, looked listless. Some of the blades were lying on their sides, tangled in a way that suggested the water had forgotten which way it was supposed to flow.

Avery reached over the side, dipping her hand into the cool, 72-degree water. It was the same temperature it had been her entire life, but the *pressure* was absent. There was no resistance against her palm. It felt like holding her hand in a bowl of still water in a suburban backyard, not the roaring, pressurized throat of a primary magnitude spring.

She took a sharp breath, the damp, rich scent of decaying cypress mulch filling her lungs. It was an earthy, comforting smell, but today it carried a hint of stagnancy. A juvenile spotted sunfish darted past the hull—too fast, too frantic. It was looking for cover that usually provided a deeper, cooler sanctuary, but the drop in water levels had exposed the limestone shelf, leaving the fish skittish and vulnerable.

It's shrinking.

The thought, once a whisper of anxiety in the back of her mind, solidified into a hard fact. Avery picked up her paddle again, her movements measured, precise. She didn't want to disturb the silt on the bottom; she wanted to see the truth. She maneuvered the kayak into the narrow, sheltered run known as the "Cypress Crown." This was her favorite stretch, a place where the canopy knit together so tightly it blocked out the Florida sun, leaving the world in a state of perpetual, moss-draped twilight.

Usually, the current here was strong enough to push her back if she stopped paddling. Now, she felt a strange, sucking sensation against the keel. The water level had receded, revealing a jagged, ghostly line of white limestone on the banks, marking where the water had sat only a week ago. That was a difference of three inches. In a spring that had been stable for as long as historical records existed, three inches wasn't a fluctuation; it was a hemorrhage.

She stopped paddling entirely, letting the kayak drift into an eddy. Silence settled over her—not the peaceful, heavy silence of the spring, but a hollow, expectant void. Even the cicadas seemed to be humming a pitch lower, their frantic energy dampened by the humidity.

Avery pulled a small, waterproof notebook from her life vest pocket. Her handwriting was small and cramped, filled with data points that nobody else had asked for: *Flow rate: sluggish. Eelgrass orientation: lateral. Surface level: -3" from benchmark. Turbidity: low, but static.*

She looked up, catching sight of a Great Blue Heron standing on a submerged log. It didn't look like a hunter today; it looked like a sentinel. Its eyes, bright and golden, tracked her movement with an unnerving intensity. Avery held her breath, trying to match the stillness of the ecosystem. She noticed the way the moss on the log above the waterline was beginning to curl, losing the moisture it desperately needed to survive. The spring was literally receding from its own life support.

She knew she shouldn't be feeling this. She was just a girl who liked the water, a student who worked the rentals. But the spring was a living thing to her, a singular, massive entity that breathed for the entire basin. When it breathed, she breathed. And right now, the air felt thin.

She turned the kayak, her stroke stronger this time, fighting the urge to sprint back to the dock. The rhythm of the water wasn't just off; it was *wrong*. It felt like a clock losing seconds. Every stroke toward the main basin felt like she was paddling away from safety and toward a disaster that everyone else was too busy to notice.

As she rounded the final bend, the main basin of Glasswater came into view. The tourist platform was already being prepped for the Discovery Day crowds. The bright, flapping banners of the park staff fluttered in the breeze, vibrant and oblivious. The water in the center of the spring—the bright turquoise eye that gave the park its name—looked shallow, the pale sand of the bottom reflecting the sunlight with an intensity that hurt her eyes. It was too bright. It was exposed.

Avery slowed, her paddle dipping into the water, feeling the lack of depth even here, where it should have been deepest. She watched a rental boat full of laughing tourists pass her, their wake slapping against the now-exposed cypress knees. They didn't notice the way the water level clawed at the roots, desperate and retreating.

She gripped the paddle until her knuckles turned white. The rhythm of the spring was broken, and she was the only one holding the metronome. She took a deep breath, the taste of ozone and pine needles sharp on her tongue, and pushed her kayak toward the dock, the urgency now a burning weight in her chest. The spring wasn't just drying up; it was crying for help, and she was the only one who spoke its language.

Avery hopped onto the weathered cedar planks of the rental dock, the wood slick with morning dew. She hauled the bow of her kayak up, the plastic keel scraping against the dock with a sound that seemed, in the quiet morning, like a scream.

The Discovery Day buzz was already hummed into the air. A cluster of brightly colored folding chairs lay stacked near the main visitor kiosk. A banner, *Glasswater Spring: Celebrating Our Crystal Heart*, rippled in the heat-heavy breeze. It was a cruel irony. The heart was skipping beats, and everyone was busy decorating the casket.

She tied off the kayak's bow line with trembling fingers. She needed to tell someone. Not just anyone—not the seasonal high-schoolers who were currently struggling to inflate a decorative archway—but someone who would look past the surface ripples and see the hollow void beneath.

She spotted Micah near the trailhead. He was kneeling, his hands buried deep in the soil of a native plant display, carefully patting down mulch around a cluster of swamp milkweed. He looked, as always, like he belonged to the earth itself.

Avery crossed the grass, her boots heavy. "Micah?"

He looked up, wiping his forehead with the back of his wrist. A wide, easy grin spread across his face, but it dimmed the moment he met her eyes. He stood, brushing off his knees. "Avery? You're early. The first group isn't here for another hour."

"The spring," she started, her voice sounding thin against the mounting noise of park rangers driving utility carts past them. "It's not just low, Micah. It's wrong. The flow—it's choking."

Micah frowned, his brow furrowing as he looked toward the water. "It's been a dry month, yeah. The seasonal ebb is always tough on the eelgrass."

"No," she insisted, taking his arm, pulling him toward the edge of the boardwalk. "Look at the cypress knees. They're entirely exposed. Yesterday, the water line was an inch higher. I've lived on this run my whole life, and I have never seen the current drag this sluggishly. It's like something is pulling the plug from the bottom."

Micah walked to the railing. He stared at the water for a long moment, his silence thick with genuine concern. "The turbidity is high, too. Usually, this time of morning, you can see down to the vents. Now it's... murky. Almost like sediment is being kicked up from somewhere deeper."

"Exactly," Avery said, feeling a wave of relief that at least one person wasn't dismissing her. "And the wildlife—the bass are skittish. They're clustering in the shadows of the bank, refusing to swim the main channel. Everything feels like it's waiting for a collapse."

"Wait," Micah said, his eyes scanning the path behind her. "Did you see Declan? He was by the dive shop earlier. If the flow is being altered at the conduit level, he's the one who'll know."

Avery nodded. "I'll find him. But Micah, we have to stop the tours. If the water drops any further, those boats are going to bottom out and tear up the floor—what's left of it."

"I'll talk to the Ranger," Micah said, his tone shifting from casual naturalist to urgent protector. "Go find Declan. If you're right, this isn't a drought. This is an extraction."

Avery left him and cut across the manicured lawn toward the research station. She found Declan hunkered over a topographic map spread across a folding table, his drone remote sitting beside a thermos of coffee. He was muttering to himself, tracing a line on the paper with a highlighter.

"Declan," she said, breathless.

He didn't look up, his mind clearly miles beneath the surface. "Avery, look at this. These original survey maps from the fifties? The drainage patterns for the north ridge were supposed to be redirected away from the limestone basin during the last expansion, but look at the grade. If there were a sinkhole collapse in the upper quadrant, the whole runoff system would invert."

"The spring is dying, Declan. Right now. The flow has dropped three inches since dawn."

That made him look up. His eyes, usually sharp with the abstract joy of geological puzzles, went cold. He stood up, knocking his chair back. "Three inches? In four hours? That's not a drought. That's a rupture."

"That's what I said," Avery replied, glancing toward the sound of an approaching megaphone. The tourists were arriving. The laughter of families, the chatter of children—it felt like a discordant soundtrack to a disaster.

A tall woman in a crisp, pressed park shirt—Zara—was walking toward them, clutching a digital voice recorder. She looked like she'd been up all night, her blazer slightly wrinkled, her eyes bright with the hunt. She had been covering the park's management issues for the local paper for weeks, and she seemed to sense the tension in the air like a shark smells blood.

"Something's happening," Zara said, bypassing the greeting. She pointed her recorder toward the water. "The park manager just issued a 'water conservation' advisory for the event, but he's calling it a 'natural seasonal fluctuation.' Does that track with what you two are seeing?"

Avery stepped forward, her voice finding its steel. "It's not natural, Zara. And it's not just a fluctuation. If we don't look at the source right now, there won't be a Glasswater Spring to celebrate by sunset."

Declan grabbed his drone bag, his pragmatism taking the reins. "Avery, you keep eyes on the basin. Zara, you get the manager on the record—ask him specifically about the new pipe installations near the north drainage ditch. I'm going to fly the basin. I want to see if the intake velocity has changed at the primary vent."

"And the community?" Zara asked, her pen hovering over her notebook. "If I report this as a disaster, we might just start a panic."

"If we don't report it," Avery countered, looking at the water—now visibly darker, the emerald hue fading into a dull, muddy brown—"we're just witnesses to the end."

The sound of an approaching bus signaled the arrival of the first wave of Discovery Day participants. The music started—a cheery, acoustic track meant to evoke the 'pristine Florida wilderness.' Avery felt a surge of indignation. She looked at her friends—the naturalist, the geologist, the journalist. The quiet observer was done waiting.

"Start the investigation," Avery said, turning toward the dock where her kayak waited. "I'm going out there. I'm going to follow the current to where it stops."

She pushed off the dock, the kayak sliding into the water with a heavy, unnatural thud. The water level was even lower than before. She looked up at the boardwalk, where hundreds of people were beginning to gather, unaware that the ground they stood on was effectively becoming a sieve.

The adventure hadn't been planned; it had been forced upon them by the spring's slow, rhythmic heartbeat as it began to stutter. Avery paddled into the center of the run, the water swirling around her paddle in lazy, suffocated eddies. She could feel the pulse of the earth beneath her, and for the first time, it didn't feel like home. It felt like a warning.

Chapter Two

CURRENTS OF CYPRESS CROWN

The mist clung to the surface of the Glasswater like a ragged, cooling blanket. At this hour—just as the first charcoal streaks of dawn bled into a pale, bruised violet—the spring usually exhaled. It was a rhythmic, lung-like expansion of water, pushing out from the aquifer with a cool, steady pressure that Avery had known since she was six years old.

She sat in the center of her kayak, the paddle resting across her lap like a dormant wing. She wasn't moving. She was listening.

Usually, the current against the hull of her boat made a specific, low-frequency hum—a liquid cello note that vibrated through the polyethylene plastic and up into her spine. It was the sound of the earth breathing. But today, the note was fractured. The steady pulse was jittery, punctuated by strange, rhythmic sighs of air caught in the limestone conduits below.

Avery leaned forward, her brow furrowing. She dipped her hand into the water. It felt thinner, lacking the thick, velvety viscosity of the spring's mineral-rich water. It felt like tap water—sterile and hollow. She drew her hand back, droplets clinging to her skin, and looked toward the "Cypress Crown," a cluster of ancient, knee-knotted trees that guarded the deepest vent of the spring.

She paddled forward, the stroke of her blade silent, honed by years of practice. As she reached the threshold of the deep-water basin, the water color changed. It wasn't the signature, crystalline turquoise of Glasswater. It was darkening, clouding with a faint, silty suspension that shouldn't have been there.

She drifted over the primary vent. Beneath her, the limestone throat of the spring usually churned with a swirling, turbulent clarity. Now, it looked like a spent lung. The water was receding, retreating into the subterranean dark, creating a visible vortex that sucked at the surface. It was a vacuum—a desperate, unnatural thirst.

Avery grabbed her camera, her fingers trembling slightly as she adjusted the aperture. She snapped a series of frames, the shutter click sounding painfully loud in the morning silence. She wasn't just looking at a seasonal low; she was looking at an extraction. Something was pulling the water out from beneath them, and the spring was dying in real-time.

She felt a sudden, piercing cold that had nothing to do with the air temperature. It was the realization that the sanctuary she had spent her life protecting was not just changing—it was being stolen. The serenity of her morning rituals evaporated, replaced by a cold, hard knot of dread in her stomach. She turned her kayak around, her strokes sharp and aggressive, cutting through the water with a frantic energy that felt alien to her.

She didn't head for the main boat ramp. She steered toward the shadow of the cypress roots where she knew Micah would be conducting his sunrise bird census.

As she rounded the bend, the familiar sights of the shoreline began to reveal their own testimony of the decline. The edges of the bank, normally submerged and lush, were exposed. Muddy, root-tangled shelves of earth lay gasping in the dawn light.

Micah was there, standing knee-deep in the shallows, his binoculars dangling around his neck. He was looking at a patch of river moss that had turned a sickly, translucent grey—a color it should never be.

"Micah," she called out, her voice cracking.

He turned, his usual calm expression tightening into a mask of sharp concern the moment he saw her face. "Avery? You're out early. What is it?"

She pulled her kayak against a cypress knee and stepped out, the mud squelching beneath her water shoes. She didn't offer a greeting. She grabbed his arm, pointing toward the bank. "Look at the moss. Look at the water line."

Micah knelt, his hand hovering over the brittle, dry vegetation. He touched it, and the moss crumbled into dust between his fingers. He looked up, his eyes meeting hers. "It's desiccated. This species requires constant saturation. It's not just a drop in the tide, Avery. This is a sustained loss of hydration."

"It's the vent," she said, her voice dropping to a whisper. "I was over the primary channel. It's pulling air. The pressure is completely gone, Micah. It's not a drought. I've seen droughts here—the spring slows, but it doesn't *collapse* like this. This is a mechanical draw."

Micah stood up, his gaze scanning the horizon, then shifting to the darkening forest canopy. "The Great Blue Heron pair that nested on the north bank... they're gone. And the manatees? They didn't even crest the run this morning. They know. The whole ecosystem is pulling back from the center."

He looked at her, and for the first time, she saw the peacemaker in him vanish, replaced by the eyes of a naturalist who realized he was watching a murder. "If the water stays this low, the salinity levels are going to spike. The plants won't just dry out; the saltwater intrusion from the coast'll poison them. Everything here is adapted to this exact flow. If the flow breaks, the chain breaks."

Avery felt a surge of adrenaline, a sharp, cold focus that pushed aside her fear. "We can't just watch it happen. The ranger thinks it's seasonal. He's telling the tourists everything is fine."

"The tourists aren't seeing what we're seeing," Micah said, his voice hardening. "They see the surface. We see the heartbeat. We have to prove it, Avery. Not with anecdotes, but with the map of the failure."

"Declan," Avery said, the name hitting them both like a rallying cry. "He has the drone footage of the ridge. He's been mapping the geological fault lines near the old development sites."

"And Zara," Micah added, his hands balling into fists at his sides. "She's been digging into the permit records for the new irrigation updates on the perimeter farms. If we link the surge in agricultural pumping to this sudden collapse in the conduit, we have a case."

Avery looked out at the spring, which seemed to groan as another air bubble burped to the surface, rippling the center of the pool. "We stop being observers today," she said, her voice steadying. "We start being guardians."

Micah nodded, his empathy now channeled into a singular, sharp resolve. "Find Declan. I'll meet you both at the boardwalk. We have to synthesize what you saw in the vent with what the data is showing us. If we don't have an answer before the park opens at nine, the flow might be gone before we can prove who killed it."

Avery nodded, pushing her kayak back into the shrinking water. As she climbed in, she took one last look at the Cypress Crown. The trees looked like silent witnesses, their roots exposed and vulnerable. She wasn't just a girl with a kayak anymore; she was the first line of defense for a world that was counting on her to notice exactly when, and how, the heart stopped beating.

She paddled away, not looking back, her mind already mapping the path to the ridge, to the data, and to the truth that lay waiting in the hollowed-out depths of the spring.

The kayak scraped against the exposed limestone shelf, a sound like grinding teeth that made Avery wince. She didn't paddle; she let the water—or what was left of it—carry her. The current here, usually a brisk, crystal-clear ribbon that pulled at the hull with welcoming intent, felt sluggish. It was viscous, heavy with sediment that shouldn't be suspended in the water column this early in the morning.

She dipped her hand into the flow. It was warmer than it should have been. *Too warm.*

Avery maneuvered the kayak into the shade of a massive, ancient cypress, its knees knobby and exposed, sticking out of the mud like the bleached bones of a prehistoric creature. Micah was already there, kneeling in the muck, his fingers gently brushing the surface of a patch of river moss that had turned a sickly, translucent gray.

He looked up as she drifted closer. His face, usually defined by a calm, observant warmth, was tight, shadowed by an anxiety that mirrored her own.

"It's not just the level, Avery," Micah said, his voice barely rising above the rhythmic *plip-plop* of a drying feeder stream. "The PH balance is spiking. I checked the test strip near the grotto."

Avery pulled the kayak's bow onto a patch of damp sand, stepping out with a grace born of a thousand hours on the water. She didn't bother with pleasantries; they were past that. "I felt the drag back near the bend. The water isn't just low—it's stagnant. It's like the spring is inhaling instead of exhaling."

"Look at this," Micah pointed to the moss. "This is *Fontinalis*. It's a cold-water indicator. It should be lush, deep green, thriving in the recharge flow. It's dying, Avery. It's dehydrating in its own home."

Avery knelt beside him. She watched a small sunfish hover near a limestone fissure, its gills fluttering with a desperate, frantic rhythm. The fish wasn't feeding. It was struggling to find enough oxygen in the thickening, warming water. She looked up at the sunlight filtering through the canopy; the spring run, which usually sparkled with the vibrancy of a living thing, now held the hazy, muted quality of a graveyard.

"I tried to trace the pressure back," Avery said, her voice dropping to a low, intense register. "There's a suction vortex near the lower vent. It's not just that the water is gone. Something is pulling it away from the other side. Someone is drinking from our well, and they're taking more than the aquifer can replenish."

Micah stood up, wiping his muddy hands on his shorts. "I saw a crew near the northern boundary last week. They were installing new high-capacity irrigation pumps for that luxury subdivision, the one they call 'The Enclave.' I thought they had the proper permits. I thought the park rangers were monitoring the flow meters."

"The rangers think it's a drought," Avery countered, standing to face him. "They're telling tourists the water is low because of the season. But look at the vegetation patterns. This isn't a natural decline. This is surgical. It's localized."

"If we don't prove it, the damage becomes permanent," Micah said, his empathy shifting into something sharper, something more protective. "The habitat won't recover if the salinity hits the tipping point. The turtles, the manatees that migrate through here, the rare grasses—they'll be gone before the end of the summer."

Avery grabbed her camera, but she didn't take a picture of the landscape. She zoomed in on the sediment, the dying moss, and the unnatural, swirling eddies of the vent. She needed evidence that couldn't be dismissed as a teenager's imagination.

"Declan and Zara are on their way," she said. "Declan has the drone. If we can map the intake velocity across the basin, we can show the difference between natural seepage and forced extraction."

"And Zara?"

"Zara is going to tear through the permit archives. If those pumps are operating outside the legal limit, she'll find the paperwork. She's the only one who knows how to make the paper trail scream loud enough for people to hear."

Micah nodded, a grim determination setting in his jaw. He reached out and touched the trunk of the cypress, a silent apology to the tree. "We aren't just observing anymore, Avery. We're guarding. If the spring dies, this whole community loses its heartbeat. People don't realize that yet, but they will."

"They will," Avery agreed, looking back toward the park entrance. She could hear the faint, growing murmur of the first tour buses arriving for Discovery Day. Families were getting off, excited to see the clear, blue water they had driven hours to witness. They had no idea they were walking into a disaster zone.

"We need to move," she said. "Before the crowds overwhelm the trails. We meet at the Limestone Grotto in twenty minutes. Bring the water samples."

"Understood," Micah said.

As they walked away from the water's edge, Avery cast one last look at the spring. The water rippled, a weak, shallow pulse, and for a terrifying second, she felt the silence beneath it—a vast, hollow pressure. The sanctuary was shrinking, but as she caught Micah's eye, she realized it wasn't just them against the drought anymore. It was them against the apathy that allowed the spring to be drained in the first place.

The investigative mission had begun, and the quiet, observant girl who once spent her mornings dreaming of photography was gone. In her place stood someone ready to fight for the current that defined her life.

Chapter Three

A Drop in the Basin

The morning sun didn't shimmer on the Glasswater; it seemed to recoil from it.

Avery pulled her kayak paddle from the water, the wood clicking softly against the hull. The sound felt too loud in the stillness of the Cypress Crown run. Usually, at six in the morning, the spring was a symphony of bubbling limestone vents and the frantic, rhythmic surface tension of schooling minnows. Today, the water was unnervingly flat, thick with a suspended, silty haze that shouldn't have been there.

Micah sat in the bow, his hands trailing through the water, his fingers brushing against a patch of *Vallisneria*. He didn't look back at her. His shoulders were hunched, the posture of someone bracing for a blow.

"The eelgrass is bleaching, Avery," Micah said, his voice barely a whisper above the gentle lap of the current. "Look at the nodes. It's not just the low level. It's the way the plant is physically retracting from the substrate. It's dying from the roots up."

Avery leaned over the side, peering into the amber depths. She had spent a thousand mornings in this basin, tracking the way light fractured through the water to illuminate the white sand vents. She knew the rhythm of this spring like her own pulse. Normally, the current would be pushing against her kayak with an insistent, cool urgency. Now, the water felt heavy, stagnant—like a lung trying to breathe in an airless room.

She pointed toward a cluster of sawgrass near the bank. "Look at the tide line on the limestone shelf. We're four inches lower than we were yesterday. That's not evaporation. You don't lose four inches of aquifer-fed water overnight to the sun."

Micah reached into his waterproof bag and pulled out a small, glass sample vial. He dipped it into the water, watching the sediment swirl like a miniature cyclone. "I checked the nitrate sensors near the northern sinkhole yesterday. The spike was already outside the seasonal norm, but this? This is different. This is a total loss of pressure."

"Pressure," Avery repeated, the word grounding her. She thought of Declan's maps—the labyrinth of limestone tunnels, the way the spring functioned as an artery for the entire region. "If the pressure is gone, it's not because the water isn't there, Micah. It's because it's being pulled somewhere else."

She grabbed her waterproof notebook, her pen scratching against the paper. She didn't write prose; she wrote coordinates and observations. *7:12 AM. Flow velocity: negligible. Turbidity: high. Flora health: Stage 2 desiccation.*

"We're not just looking at a bad season," Avery said, her voice sharpening. The quiet, observant teen who usually let the beauty of the spring speak for itself was replaced by someone who saw the Glasswater not as a park, but as a patient on a table. "This is a systemic failure. If we don't document the difference between these levels and the historical baselines before the crowds get here, it'll be impossible to prove it later."

Micah nodded, a grim resolve settling over his face. He began to mark the shoreline with bright orange surveyor's tape, not as a volunteer for the park's 'Discovery Day,' but as a sentinel. "The park rangers are going to be out here in an hour. They'll be setting up the signs for the kids, putting the kayaks out for the 'Spring Discovery' tour. They won't want to see this."

"They won't see it if they aren't looking," Avery countered. She maneuvered her kayak toward the narrow gap in the cypress knees that led toward the Grotto. "That's why we're going to be the ones looking."

They moved with an efficiency born of shared anxiety. Micah mapped the distribution of the wilting vegetation while Avery recorded the water temperature at three different depths, noting the anomalous warmth creeping up from the vent. It was unnatural. The spring was supposed to be a constant, cool seventy-two degrees, a steady heart beating against the humid Florida heat. The heat was a fever.

As they reached the edge of the Grotto, the entrance was shrouded in a thin, grey mist. The air felt thin, ionized.

"Do you feel that?" Micah asked, pausing his work. He looked toward the treeline, where the park staff's utility carts were beginning to rumble in the distance. The day was starting. The cheerfulness of the event was about to clash with the reality of the dying spring.

Avery pulled her camera from her bag, the long lens fixed on the vent. She didn't want to capture the beauty of the Glasswater this time; she wanted to capture the evidence of its struggle. She adjusted her aperture, the shutter clicking rapidly.

"They're going to tell the visitors it's just a cycle," Avery said, her eyes locked on the viewfinder. "They're going to tell them the spring is resting, or that the rain just hasn't been enough."

"And if we let them?" Micah asked, his voice thick with emotion.

Avery finally looked up, her expression unyielding. "Then the cycle stops. Permanently. We can't let them turn the page on this, Micah. We have to make the story impossible to ignore."

She tucked the camera away and dug her paddle into the mud, pulling them closer to the hidden, shadowed mouth of the Grotto. They had minutes before the park opened. They had a mountain of data, a dying ecosystem, and a growing, desperate need to be heard.

"Let's move," Avery said, her voice steady. "The Grotto is the only place we can finish this. If the foundation is collapsing, we need to know exactly how deep the hole goes."

They pushed forward, disappearing into the shade of the cypress, leaving the calm morning behind for the cold, hard work of the investigation. The Glasswater was still breathing, but for the first time in her life, Avery could hear the rattle in its throat. She wasn't just observing the spring anymore. She was fighting for it.

The humidity hung over Glasswater Spring like a damp wool blanket, heavy and suffocating, even at eight in the morning. On the boardwalk, the sounds of setup—the rhythmic clatter of folding tables, the squawk of walkie-talkies, and the nervous banter of volunteers—already signaled the start of Discovery Day.

Avery stood at the edge of the overlook, her gaze fixed on the Grotto's mouth. Below, the water wasn't the usual translucent turquoise; it was a dull, bruised slate.

"They aren't even looking down," Micah whispered, stepping up beside her. He clutched his notebook, his knuckles white. "They're too busy setting up the 'Water is Life' displays to realize the water is currently dying."

Avery pointed to a limestone shelf that should have been submerged by at least two feet. Instead, it was slick with exposed, drying algae. "Look at the *Vallisneria*. The ribbon grass is lying flat against the silt. It's not just thirsty, Micah. It's gasping."

They moved toward the staging area where Ranger Miller, a man who looked like he'd rather be anywhere else, was barking orders at a group of students hanging banners.

"Ranger Miller," Avery started, her voice steady despite the adrenaline spiking in her chest. "We need to talk about the flow rate. The spring head isn't just low—it's stagnant. The oxygen levels near the vents are crashing."

Miller wiped sweat from his forehead with a bandana, barely glancing at them. "Avery, I appreciate the enthusiasm. I really do. But it's the dry season. We've seen this flux a dozen times before. The springs are part of a larger system; sometimes it breathes in, sometimes it breathes out."

"This isn't a breath," Micah insisted, his voice rising with a rare edge of sharpness. "The aquatic vegetation is showing signs of thermal stress. If we push the foot traffic from the tours across the boardwalk today, the vibration alone could destabilize the saturated soil banks. We need to postpone the tours."

Miller let out a short, dismissive laugh, gesturing toward the growing line of families near the entrance. "Postpone? Kids are coming from three counties to learn about the aquifer, and you want me to cancel because the water's a little murky? The state hydrology report says we're within the 'acceptable fluctuation' margin. Help Zara with the press packets. That's your job today."

He turned his back, dismissing them with a flick of his wrist.

Avery felt a cold, hard knot form in her stomach. It was the same feeling she got when she saw a storm front moving in over the Gulf—a mixture of dread and clarity. She looked at Micah. His empathetic eyes had hardened into something more analytical, more focused.

"Acceptable fluctuation," Avery repeated, her voice low. "He's reading a report that's two weeks old, maybe older."

"He's reading what he's told to read," Micah replied. He looked down at his watch, then back at the boardwalk, where tourists were already beginning to swarm. "They're going to walk right over the most delicate section of the conduit. If they don't listen, we show them."

"How?"

"We don't need permission to measure the world, Avery."

They retreated to the shade of a cluster of sabal palms, well away from the rangers' line of sight. Declan was already there, hunched over his tablet, his drone—a sleek, compact piece of tech—buzzing softly on the ground. Zara stood nearby, frantically scrolling through a series of digital maps on her phone.

"Miller is a brick wall," Avery said, keeping her voice hushed.

"I gathered," Zara muttered, not looking up. "I just pulled the recent permits for the Enclave expansion project. They've been granted a 'temporary emergency increase' on their irrigation pumping rights. They claim it's for dust mitigation during construction, but the volume they're pulling? It's massive. It's enough to draw down the head pressure of a minor spring system in a matter of days."

Declan looked up, his face grim. "That explains the inversion I'm seeing. I ran the sonar mapping of the grotto a few minutes ago. The flow isn't just slowing down; it's being redirected. There's a localized vacuum forming in the secondary cave network. The spring isn't drying because of the weather. It's being siphoned."

Avery crouched, pulling a measuring tape and a digital sensor from her gear bag. "If we're going to stop this, we need to prove the correlation. Not just 'it's low,' but 'it's dropping at X rate per hour while their pumps are active.' We need raw, indisputable data."

"We're going to be the ones the park kicks out, aren't we?" Micah asked, though there was a trace of a smirk on his lips.

"If it saves the Grotto, it's worth the ban," Avery said. She looked toward the water, where the first group of kayak tourists was preparing to push off, their paddles poised to scrape against the shallow, fragile limestone. "Declan, get the drone into the canopy above the pump station. Zara, keep digging into those permits. If we can link the Enclave's pump logs to the minute-by-minute drop in the spring level, we don't just have a theory—we have a case."

"What about us?" Micah asked.

Avery grabbed her camera, the lens already adjusted for low-light macro photography. "We're going into the water. If the ecosystem is dying, we document the anatomy of the collapse. We record everything—the sediment, the temperature, the gas levels. We give the community a window into what's happening beneath the surface, whether they want to look or not."

As the park's public address system crackled to life, welcoming the families to the beauty of the spring, the four of them moved into the shadows. The transition was absolute. They were no longer volunteers or students; they were the spring's last line of defense, and the clock was no longer measured in event hours, but in inches of disappearing water.

Avery stepped into the cool, brackish shallows. The water was warmer than it should have been, a sickly, tepid temperature that told her everything she needed to know. She took a breath, centered herself, and pressed the shutter on her camera.

"Start recording," she whispered.

The investigation had begun.

Chapter Four

THE DISCOVERY DAY DISRUPTION

The sun-drenched canopy of the Glasswater Spring park hummed with the high-pitched drone of cicadas and the laughter of families navigating the boardwalk. A vibrant "Discovery Day" banner flapped lazily in the humid breeze, its colors painfully bright against the backdrop of a day that was supposed to be a triumph.

Avery stood at the edge of the lower pier, her camera dangling neglected by her side. She wasn't looking at the tourists; she was staring at the limestone shelf three feet below the waterline. Or, rather, she was staring at the fact that it was no longer underwater.

A jagged scar of wet, exposed rock glistened in the morning sun, slick with slime and stripped of the lush, swaying eelgrass that usually carpeted it. The spring, the pride of the county, wasn't just low. It was gasping.

"Avery?"

Micah stepped onto the wood, his boots silent. He was holding a handheld water-testing kit, but he wasn't looking at the test strips. He was looking at the way the current, usually a vigorous, crystal-clear surge from the aquifer's heart, had curdled into a sluggish, murky crawl.

"Look at the vent," Avery whispered, pointing toward the deep, azure abyss at the center of the pool. "The flow isn't just slow, Micah. It's reversing. Look at the silt."

Micah leaned over the railing, his face pale. Tiny suspended particles that should have been swept out toward the river were swirling back into the mouth of the vent, pulled by a faint, erratic suction. He grabbed his binoculars, scanning the treeline for the usual bubbling disturbances that signaled the subterranean surge. Nothing. The water was unnaturally, deathly still.

"The dissolved oxygen levels are plummeting," Micah said, his voice barely audible over the chatter of a nearby tour group. "If the flow doesn't recover in the next few hours, the entire lower basin... it won't just be stressed. It'll be dead by sundown."

A loud, forced cheer echoed from the main pavilion. A park ranger—a man whose uniform seemed more tailored for administrative meetings than wilderness patrol—was clapping his hands, announcing the start of the 'Hydrology and Heritage' walking tour.

"Everything is perfectly normal, folks!" the ranger's voice boomed, amplified by a megaphone. "Just a bit of a seasonal dip. The spring is a dynamic system, after all. Nature's ebb and flow."

Zara, standing ten feet away, stiffened. She snapped her notebook shut, the sharp *click* of the metal spiral biting into the humid air. She marched toward the ranger, her stride purposeful, like a shark catching a scent.

"That's a lie," Zara muttered as she passed Declan, who was hunkered down by a pile of gear.

Declan was adjusting the sensors on a portable sonar unit, his jaw set in a hard, uncompromising line. He didn't look up, but his fingers tapped a rapid, jagged rhythm against the casing of his laptop. "It's not just a lie," he countered, his voice tight. "It's a mathematical impossibility. I've been running the flow velocity logs since six this morning. We're losing volume at a rate that would require a tectonic shift or a catastrophic sinkhole breach. Seasonal flow doesn't drop forty percent in four hours without a mechanical catalyst."

"Mechanical," Avery repeated, the word chilling her more than the mist rising from the spring. "Like a pump?"

"Like a massive, industrial-grade withdrawal," Declan corrected, finally looking up. His eyes were wide, the frantic energy of a researcher who had just found the anomaly he feared most. "I tried to recalibrate the sonar, thinking it was a glitch, but the data is consistent. Someone—or something—is pulling thousands of gallons a minute from the upstream conduit. The spring isn't drying up. It's being drained."

The sound of the ranger's megaphone continued, painting a picture of tranquility for the unsuspecting visitors. *"Our beloved Glasswater is resilient, resilient as the Florida spirit!"*

"He's selling a fantasy," Zara said, joining the group, her face flushed with frustration. "I just checked the park's public logs. They aren't even recording the water levels at the vent anymore. They stopped the hourly reporting three days ago. They're hiding it, and if we don't do something before the noon surge of tourists reaches the grotto, we're going to lose the chance to document the baseline."

Avery looked back at the vent. A small school of mullet broke the surface, gasping, their scales duller than they had been an hour ago. The beauty of the place, the deep, ancient wonder she had protected since she was a child, felt like it was crumbling in real time.

"They won't listen to us," Avery said, her voice growing steady, infused with a sudden, sharp clarity. "They'll call us kids overreacting to a heatwave. We need to stop acting like guests at an event and start acting like the people who actually know this water."

"Avery's right," Micah agreed, stepping back from the railing, his empathetic calm replaced by a cold, protective resolve. "If we're going to stop this, we need proof that hits harder than a park ranger's spin. We need data that they can't brush away."

"I have the sonar mapping," Declan said, standing up and slinging his pack over his shoulder. "And the drone is charged. If I can fly it over the northern ridge, I can get thermal imagery of the ground. If there's an unauthorized pump station, those units leak heat. They always leak heat."

"And I'll get the permit history for every square mile surrounding the park," Zara added, her journalistic instinct surging. "If there's an industrial draw, there's a paper trail. You don't pull that much water without leaving a signature in the county records."

"And I'll stay here," Avery said, her gaze fixed on the darkening, swirling water at the mouth of the vent. "I'll document the biological shift minute by minute. If the vegetation starts to brown or the oxygen levels drop to zero, I'll have the photos to show exactly what this 'seasonal dip' is costing us."

Micah nodded. "I'll coordinate the water chemistry. We need to show the nitrate spikes. If they're pumping that hard, they're pulling the stagnant, contaminated water from the edges of the aquifer into the main flow. It's not just about volume—it's about toxicity."

The four of them stood in a tight circle, the festive noise of Discovery Day fading into the background, replaced by the heavy, humid silence of their mutual resolve. The park ranger's voice rose again, praising the "endless bounty" of the spring.

Avery turned away from the celebration. She walked toward the water, not as a visitor, but as an investigator. The investigation had begun. The spring was dying, and they were the only ones who knew the truth of its murder.

"It's not just a low tide," Avery said, her voice barely rising above the rhythmic *thwack* of a plastic paddle against a dock pylon. She was crouched at the edge of the boardwalk, fingers trailing in the water. "The current—it's pulling *inward*. Like the spring is trying to swallow itself."

The Discovery Day crowd behind them was a blur of neon sunscreen and screaming toddlers, completely oblivious to the fact that the lifeblood of the park was receding in real time.

Declan adjusted his glasses, his eyes glued to the tablet screen displaying the sonar feed from his submerged transducer. "She's right. Look at the flow velocity data. We're seeing a negative pressure gradient near the mouth of the vent. The water isn't just failing to push out; it's being siphoned away by something massive."

Zara pushed her sunglasses onto her head, her face pale. She'd spent the last hour trying to get Ranger Halloway to look at the water-level gauges. His response—a weary, practiced wave of his hand and a comment about 'seasonal aquifer oscillation'—had been the final straw.

"Halloway isn't going to budge," Zara said, her voice tight. "He's got a line of tourists waiting for the glass-bottom boat tour. He told me if I kept 'harassing staff with doomsday theories,' he'd pull my press pass for the rest of the season. He thinks this is just a dry spell, a little less rain this month."

Micah knelt beside Avery, his hands submerged in the mud at the waterline. He pulled them out, his palms stained with a dark, silty residue that smelled faintly of sulfur and something sharper—a chemical tang that didn't belong in a protected spring.

"It's not just the level," Micah said, his voice unusually grim. "The eelgrass is already bleaching. Look at the color. It's turning white at the base, not brown. This is osmotic shock. The salinity levels are spiking because the freshwater pressure from the aquifer is gone. If this continues for another four hours, we're looking at a localized die-off of the entire nursery population."

Avery looked up at her friends. The camaraderie of the morning had vanished, replaced by a cold, sharp-edged clarity. She had spent her entire life in this basin. She knew the way the light hit the limestone, the way the turtles migrated to the deeper, cooler pockets when the summer heat peaked. This wasn't the sun or the weather. This was a theft.

"We stop waiting for them to notice," Avery said, standing up. She wiped her hands on her cargo shorts. "Declan, how much battery do you have left on the drone?"

"Enough for one deep-run, maybe two," he said, tapping the screen. "But if I fly it over the western perimeter, past the property line of the old citrus groves, I can get a look at those pump stations. If someone is pushing a high-capacity well into the Floridan Aquifer, the infrared signature will be undeniable."

"And I'll go back to the archives," Zara said, pulling her notebook from her bag. "I've got the permit logs for the 'Enclave' development project. If they're pulling more than the permitted cubic footage, there has to be a physical record of the infrastructure installed near the boundary line."

Micah stood, dusting off his knees. "I'll start mapping the toxicity spread. I need to document the water chemistry at the vent every thirty minutes. If I can prove the nitrate and salinity levels are rising in direct correlation to their extraction cycles, the science won't be ignorable."

Avery felt a tremor in her hands, but it wasn't fear—it was adrenaline. "I'll take the kayak out to the secondary vent. It's the closest point to the Enclave's underground pipeline. If the suction is happening there, I'll feel it through the hull. I'll be your eyes on the ground—or, well, on the water."

"Avery," Declan started, hesitation flickering across his face. "If that vent is siphoning as hard as the sonar suggests, that current could pull a kayak under. You know the history of those underwater cave systems."

Avery met his eyes, steady and unflinching. "I know these currents better than anyone, Declan. I'm not going *in*. I'm going to stay on the surface and track the eddy patterns. I'll keep the beacon on."

The four of them exchanged a silent look. The park's loudspeakers crackled with the tinny sound of a scavenger hunt announcement, a stark contrast to the gravity of their pact. They were no longer just teenagers exploring the park; they were witnesses to a crime against the land.

"We reconvene at the Grotto at sunset," Zara said, her journalist's instincts overriding the last of her nerves. "Whatever we find, we document. Every photo, every data point, every interview. We don't bring it to Halloway. We bring it to the local board, the water management district, and if we have to, the press."

"Let's move," Micah said, turning toward the trailhead.

Avery grabbed her paddle. The air felt heavy, saturated with the humidity that usually promised a thunderstorm, but the sky was clear. The humidity was coming from the drying basin, the soul of the spring gasping for air. As she pushed her kayak into the shallow, receding water, the boat scraped against the limestone shelf—a sound that felt like a scream.

She pushed off, the cold water swirling around her ankles. She didn't look back at the crowds. Her focus was entirely on the dark, receding mouth of the secondary vent ahead, the place where the truth lay hidden in the deep.

Chapter Five

Measuring the Void

The humidity hung over the basin like a wet wool blanket, but Avery didn't feel it. She felt the silence. The rhythmic *slap-slap* of water against limestone, which usually served as the heartbeat of Glasswater Spring, had been replaced by a discordant, hollow gurgle.

Avery pulled her kayak onto the shelf of the secondary vent—a jagged opening beneath the limestone overhang that she'd dubbed "The Throat." Normally, this spot was submerged under two feet of crystal-clear, swirling water. Now, the dark mouth of the cavern was exposed, slick with dying moss.

She heard the low hum of a drone's rotors before she saw it. A sleek, quad-copter shape crested the cypress canopy and descended with military precision, hovering just inches above the water's surface near her position. Declan was perched on a ridge two hundred yards to the north, his eyes glued to a tablet.

Avery didn't wait for him to initiate. She pulled a handheld flow meter and a collapsible surveying rod from her dry bag. She waded into the muck, the water only reaching her ankles where it should have been at her waist. The smell of sulfur and decaying organic matter—the scent of a spring choking—stung her nostrils.

"Declan," she whispered into her radio, her voice tight. "I'm at the vent. The water level is down by twenty-six inches from the morning mark. It's dropping faster than the main basin."

"Copy that, Avery," Declan's voice crackled, focused and devoid of his usual playful energy. "I'm seeing the thermal signature on the drone now, but it's erratic. The water temperature is spiking near the north pipe. If this were a drought, the heat signature would be uniform. This is concentrated. It's artificial."

Avery knelt, pressing the flow-meter against the limestone edge. The water, which should have been surging outward, felt sluggish, almost viscous. "It's not just the level, Dec. The current is reversing. The spring isn't pushing out anymore. It's being pulled."

She moved her rod, marking the exact depth against a notch in the rock—a marker she had scratched into the stone years ago during a dry summer. It was three inches below her lowest recorded mark.

"I'm engaging the sonar depth sounder," Declan said. "Avery, keep that rod steady. I need a correlation between the drone's overhead mapping and your physical immersion depth. If I can prove the depression in the water table isn't radiating from the center of the aquifer but is localized to this specific conduit, we have our 'smoking gun.'"

Avery shivered, despite the ninety-degree heat. She watched a small, translucent shrimp—a species she had studied for her biology project—struggle in a shrinking puddle near her boot. She scooped it up with her hand and placed it into a sample jar filled with the last of the deep-flow water.

"I've got the rod set," she said, her voice steadying as she settled into the rhythm of the work. "The vent is pulsing. Every time the hum from your drone gets louder, I feel a vibration in the rock beneath my boots. It's like a heartbeat, but it's mechanical."

"That's the pump-start," Declan muttered, more to himself than her. "They're cycling the irrigation system. They think the park is empty because of the Discovery Day chaos, so they're opening the valves to get the best pressure before the peak heat. They're stealing the recharge."

Avery shifted the rod. She didn't look at the sky or the park; she looked at the water. She watched the way the sediment moved—not in the natural, spiraling dance of a spring-fed system, but in straight, panicked lines toward the hidden depths of the pipe.

"I need you to mark the exact GPS coordinate of that vibration, Avery," Declan commanded.

She pulled her handheld GPS unit from her vest, squinting at the screen. "Got it. I'm tagging the site of the secondary vent. The vibration is centered exactly three meters from the primary limestone fissure."

"Perfect. Keep that position. I'm dropping a probe marker from the drone."

A soft *plink* sounded as a small, biodegradable marker hit the water near the vent. Avery grabbed it and secured it to the rock. She looked up, finally seeing Declan on the ridge. He was standing, arms braced against a cypress trunk, his eyes wide as he looked at the data scrolling across his screen.

"It's not just the Enclave, Avery," Declan said, his voice dropping into a register of profound, terrifying realization. "The map is projecting a pull from three separate sectors. They aren't just pumping from one well. They've tapped into the entire subterranean river system. If I overlay this with the old state topography maps Zara found, we're looking at a catastrophic redirection of the main karst flow. They're literally siphoning the breath out of the spring."

Avery looked at the empty, gasping mouth of the vent. The wonder she usually felt here—the feeling of being in a cathedral of water—was gone.

In its place was the cold, hard weight of responsibility. She wasn't just an observer anymore; she was a witness to an autopsy.

"How much time do we have before it stops entirely?" she asked.

There was a long silence on the radio.

"At this rate of extraction?" Declan finally replied, his voice barely a whisper. "Maybe six hours. By tomorrow morning, the spring won't just be low, Avery. It will be a mud pit."

Avery tucked her equipment away, her movements sharp and efficient. The quiet teen who spent her weekends photographing kingfishers was gone. In her place was a researcher who understood exactly what was at stake. She looked at the water, then at the drone buzzing overhead.

"Get the data to Zara and Micah," she said, her voice hard as the limestone she stood upon. "We don't need a presentation anymore. We need a stop-work order. And we need it before the sun goes down."

She shoved off from the bank, the kayak scraping against the now-exposed rock. She didn't paddle gently. Every stroke was a protest, a desperate attempt to move faster than the mechanical vacuum clawing at the heart of her home. As she cut through the glassy, dying water, she knew this wasn't about data points or geological surveys anymore. It was about saving the living, breathing life-force that had nurtured their town for a century, and the realization that the only thing keeping it alive right now was the very data they had just managed to record.

She hit the center of the basin, the water swirling around her paddle in a pale, weak imitation of its former power. The spring was crying out, and for the first time, she finally understood what it was trying to tell her. It wasn't dying of a drought; it was being bled dry. And she knew exactly who held the knife.

The Grotto was a pocket of cool, humid silence, a stark contrast to the stifling humidity pressing down on the rest of the park. Avery knelt on the mossy limestone ledge, her fingers tracing the "High Water" notch she had carved into the rock three years ago. It was bone-dry, exposed to the humid air like a scabbed-over wound.

Declan hovered over his tablet, the screen casting a pale, sterile glow against the damp cave walls. The hum of his sonar unit was the only sound besides the rhythmic, sluggish drip of water from the ceiling—a sound that should have been a steady, rushing cascade.

"Look at the bounce-back, Avery," Declan said, his voice tight. He zoomed in on a waveform that looked like a flatline compared to the jagged, healthy heartbeat of historical data. "If this were a natural drought, we'd see a tapering—a slow, geological adjustment. But this? This is an abrupt truncation."

Avery leaned in, her eyes scanning the data. She didn't need the graph to tell her the spring was dying; she could feel it in the stillness of the air, the lack of that familiar, pulsing vibration beneath her boots. "The discharge isn't just slow," she whispered. "It's being throttled. Like someone clamped a hose."

"That's exactly what it looks like," Declan muttered, tapping a command that overlaid the current sonar map onto a decades-old USGS topographical survey.

The two maps overlapped with chilling precision. Declan pointed to a series of blue-shaded subsurface conduits that snaked away from the main basin. "See these feeder lines? Under normal pressure, they should be pushing five thousand gallons per minute into the main vent. Right now, they're barely trickling. And it's not an even drop across the board."

Avery frowned, moving her hand to point at a specific knot of data points near the north edge of the map. "Wait. Here. Why is the pressure drop so much more severe on the eastern secondary vent?"

Declan squinted, his fingers dancing over the touch screen. "That's the section closest to the Enclave's new expansion project. If the groundwater were receding due to a drought, the drop-off should be symmetrical. But the eastern conduits are effectively vacuum-sealed. It's localized suction."

Avery stood up, her legs stiff from crouching. She looked up toward the narrow ceiling opening, where a sliver of twilight was beginning to bleed through. "So it's not the rain. It's not the humidity. Someone isn't just using the water; they're pulling it directly from the throat of the spring."

"It's a siphon," Declan said, his tone shifting from clinical to incredulous. "They've sunk high-capacity deep-well pumps somewhere along the northern boundary of the development. They're tapping the aquifer before it ever reaches the spring's natural basin."

Avery felt a cold prickle of anger beneath her skin. "They're killing the pulse of the park to keep their lawns green."

"It's worse," Declan corrected, his brow furrowing as he pulled up the drone telemetry from earlier that afternoon. "Look at the thermal signature I caught near the pump house at the Enclave's edge."

He pointed to a blotch of unnatural heat on the drone footage. "That's not just a pump, Avery. That's an industrial-grade draw, running continuously. It's creating a localized sinkhole effect. The reason the spring is drying up isn't just extraction—it's that they've effectively reversed the natural drainage flow. They've turned our spring into a secondary sump."

Avery paced the small ledge, the gravel crunching under her sneakers. She thought about the wildlife—the limpkins she'd seen struggling in the muddy shallows, the way the eelgrass was already turning a sickly, bleached white. "We have to prove it. Not just that the water is gone, but that their specific pumps are the ones pulling the plug."

"I have the drone footage of the pump station, and the sonar proves the flow is being diverted northward," Declan said, his voice gaining a hard, focused edge. "But we need the 'smoking gun'—the log that says they're running at max capacity during a drought restriction period. We have the 'what' and the 'where.' We need the 'why' and the 'who'."

Avery stopped pacing. She looked at him, her quiet nature replaced by a sharp, protective intensity. "Zara has the permit history. Micah is checking the nitrogen spikes. If we match their pumping hours to the chemical contamination Micah's seeing in the runoff, we won't just have a theory. We'll have a map of their negligence."

Declan nodded, packing his tablet into his waterproof bag with methodical efficiency. "We're not just kids in a park anymore, are we?"

"No," Avery said, grabbing her headlamp. She clicked it on, the beam cutting through the gloom of the Grotto. "We're the only ones keeping track of the heartbeat. And it's time we showed them exactly how fast it's failing."

As they emerged from the cave, the Florida air was thick with the scent of pine and oncoming rain—a cruel irony, Avery thought. The sky could open up and soak the earth, but it wouldn't reach the aquifer deep below, not while those pumps were running.

"We meet the others at the park entrance?" Declan asked, his silhouette stark against the deep indigo of the twilight sky.

"The boardwalk," Avery replied, already moving toward the trees. "It's where the silence is loudest. That's where we finish this."

They walked in silence for a moment, the gravity of their findings hanging between them. The forest was uncharacteristically quiet—no chorus of tree frogs, no rustle of nocturnal movement. The spring was the lifeblood of this entire ecosystem, and it was being bled dry. Avery gripped the strap of her camera bag, her knuckles white. She had spent her life documenting the beauty of this place; now, she was documenting its expiration.

"Declan?" she said, pausing at the edge of the clearing.

"Yeah?"

"When we present this... when we show them what's happening... people are going to be angry. The Enclave, the developers, maybe even some of the neighbors who work there."

Declan checked his drone remote, ensuring the SD card was secure. "Let them be angry. I'd rather face an angry board of directors than a dead spring. Wouldn't you?"

Avery looked back toward the dark basin of Glasswater. The water was now a stagnant, mirror-like surface, reflecting the stars—a tombstone for a miracle. "Yeah," she said softly. "I would."

They started walking again, picking up the pace, the urgency of the discovery driving them toward the meeting point. The mystery wasn't a puzzle anymore; it was a countdown. And they were the only ones holding the clock.

Chapter Six

Whispers of the Aquifer

The Grotto was a place of perpetual twilight, a limestone cathedral where the air tasted of wet stone and ancient minerals. Avery adjusted the strap of her waterproof gear bag, her fingers brushing against the cold, jagged edge of the canyon wall. Beside her, Declan was hunched over a sonar device, the screen casting a sickly blue glow against the glistening white rock.

"The pressure differential is worse than yesterday, Avery," Declan whispered, his voice barely audible over the rhythmic drip-drip of water from the ceiling. "It's not just lower. It's... stalling. Like a heart with a blockage."

Avery knelt at the edge of the pool. She watched the current. Normally, the water at this depth danced with a frantic, crystalline energy, pushing debris toward the mouth of the spring run with purpose. Now, it moved with a sluggish, oily lethargy. A strand of eelgrass, once a vibrant, swaying emerald, was limp, coated in a fine, grayish silt that shouldn't have been there.

She reached out, her hand breaking the surface. The water felt tepid—flat, lacking the characteristic biting chill of the Floridan Aquifer.

"The filtration pulse," she murmured, pulling her hand back. "It's gone."

Declan looked up, his expression sharpening. "What do you mean?"

"The spring breathes," Avery said, staring into the dark, receding water. "It takes in groundwater and exhales it through these conduits. It's a rhythmic pressure release. My grandfather used to say if the spring stops breathing, the forest starts to die. I thought it was just a local legend—a story to keep kids from playing in the intake tunnels. But look at the silt, Dec. The current isn't strong enough to keep the passage clear anymore. We're watching the plumbing of the entire ecosystem seize up."

Declan moved beside her, his drone remote hanging heavy at his hip. He looked at the sonar data—a jagged, stuttering line that showed none of the fluid, rolling waves typical of a healthy spring system. He tapped the screen, trying to recalibrate, but the data remained flat.

"If the pressure is gone, it means something is pulling on the other end," Declan said, his voice hardening into a pragmatic edge. "We aren't just watching a drought. We're watching an extraction. Something is vacuuming the aquifer dry from the outside."

"The Enclave," Avery said, the name hitting the damp air like a stone.

"We need to know how deep it goes," Declan replied, packing the sonar with sudden, precise movements. "If I can get a sensor into the main conduit—the primary feeder—we can prove the flow rate is being manipulated. Not by weather, not by climate, but by pumps."

"That's restricted territory, Declan. The park management blocked the northern tunnel two years ago for 'structural safety.'"

Declan looked at her, his eyes reflecting the blue light of his tablet. "If we don't go now, there won't be a spring left to protect. You've felt it, Avery. You've felt the spring dying since Discovery Day. Do you want to be the one who documented the end, or the one who stopped it?"

Avery took a breath, the damp air filling her lungs. She looked toward the narrowing throat of the Grotto, where the water tunneled into the absolute dark. The explorer in her—the girl who knew every current—was terrified. But the protector in her was furious.

"Give me the sensor," she said, her voice steady.

They moved into the tunnel. The transition was jarring; the ambient light of the Grotto vanished, replaced by the narrow beam of their headlamps. The walls here were different—less natural, more eroded, showing the scars of a shifting water table. They waded through waist-deep water that felt wrong against their skin—heavy, chemical, and strangely warm.

Avery led the way, her hands finding the familiar grooves of the rock. She moved by memory, her body adjusting to the nuances of the cavern floor. Every few feet, she stopped, listening. There was a sound beneath the silence—a low, rhythmic thrumming that vibrated through the soles of her boots. It wasn't the sound of water. It was the sound of machinery.

"Do you hear that?" she whispered.

Declan froze, holding his breath. "That's a pump. A big one."

He gestured to a small, jagged alcove in the limestone. Together, they pried a heavy, waterproof canister containing their makeshift flow-monitor into a shelf in the rock. As they secured it, Avery shone her light on the cavern wall. It wasn't just limestone. There were patches of concrete and rusted rebar embedded in the rock—remnants of a forgotten drainage project, likely from the old housing site developers had abandoned decades ago.

"Look at this," she pointed to a vein of white, chalky residue streaking the stone. "That's not a mineral deposit. That's runoff. Fertilizer, pesticides—stuff that shouldn't be here."

Declan leaned in, his face tight with anger. "They aren't just pumping the water out. They're allowing the surface water to leak directly into the aquifer because they ignored the drainage standards. They've cracked the limestone cap."

The realization hit them both with the weight of a physical blow. This wasn't a natural disaster. It was a slow, calculated theft of the water, compounded by decades of corner-cutting and apathy. The spring wasn't just losing its flow; it was being poisoned from within.

Avery looked back toward the Grotto. "We have enough. We have the data, and we have the physical evidence of the contamination."

Declan nodded, his hands trembling slightly as he secured the final cable. "We need the others. If we go public with this, we're going to be up against people with a lot more influence than a park ranger. We need to be airtight."

They waded back toward the Grotto, the thrumming of the distant pump growing fainter but never disappearing. As they emerged into the dim light of the main chamber, the reality of their situation settled over them. They were no longer just curious teens exploring a park. They were witnesses to a crime against the land.

"The boardwalk," Avery said, her voice echoing off the cavern walls. "We rendezvous with Zara and Micah at the shed. We show them everything."

"And then?" Declan asked, shielding his eyes as they moved toward the exit.

Avery looked at the struggling, shallow water of the spring run, seeing it not as a static feature, but as a wounded, living thing fighting for its life.

"And then," she said, "we give the spring a voice."

The abandoned research shed smelled of mildewed paper and wet limestone. It was a relic of a time when the park took science seriously, now serving only as a graveyard for rusted calipers and water-damaged surveys.

Avery wiped a layer of grime from the central drafting table, revealing a topographical map of the Glasswater basin from 1974. Beside it, the digital tablet Declan held pulsed with the blue light of the new, terrifying data they had just pulled from the upstream conduits.

"Look at the pressure drop," Declan said, his voice tight. He tapped the screen, dragging a jagged graph over the faded ink of the old map. "The sensor tripped at 0300 hours. The flow rate isn't just fluctuating—it's hitting a hard mechanical ceiling. It's cycling, almost rhythmically."

Zara paced the narrow aisle between the desks, her phone recording a voice memo as she paced. "Cycling? Like a pump?"

"Exactly," Declan replied. "An industrial-grade extraction pump. It's pulling from the primary conduit, effectively stealing the pulse before it even reaches the main basin."

Micah stood by the window, watching the moonlight reflect off the stagnant, muddy surface of the spring. He turned, holding a vial of water he'd collected earlier. It was cloudy, a far cry from the crystal-clear turquoise that had defined his childhood. "I ran the field test on the nitrates, and I cross-referenced it with the sediment levels we saw near the intake pipe. This isn't just diversion, guys. The water being pumped out is contaminated with high concentrations of phosphate and synthetic fertilizer. They're sucking the spring dry, then dumping the runoff from that housing development—The Enclave—straight into the upper reaches of the aquifer."

Avery leaned over the table, her fingers tracing the contour lines of the 1974 map. She looked at the location of the modern development, then at the natural limestone fractures marked in red ink on the antique survey. "It's not just a bad decision by a contractor. Look here." She pointed to a series of dots along the northern ridge. "The old maps show this entire ridge was a natural filtration swamp. It was designed to recharge the aquifer slowly. But in the satellite overlay, all of this is concrete. They paved the sponge."

Zara stopped pacing and crowded in, looking at the contrast between the old, lush geography and the current, hardened reality. "So, the spring isn't just drying because they're pumping water out. It's drying because the rain has nowhere to go. It's running off the pavement, picking up toxins, and bypassing the natural filtration system entirely."

"It's a double hit," Micah added, his voice low and heavy. "The pump creates a vacuum that keeps the spring from recovering, and the paving prevents the replenishment. We're watching the entire cycle get strangled to death."

The silence in the shed was thick, broken only by the hum of the cooling fan on Declan's tablet. The weight of the discovery—the sheer, systemic scale of the destruction—hung over them. It wasn't just a mystery anymore; it was an indictment of how the town had grown, one permit and one parking lot at a time.

Avery pulled a notebook from her bag, flipping past sketches of heron and egret to a blank page. She didn't look up. "We need to connect the dots. The permits Zara found—they were signed off by the County Planning Commission, right?"

"Twenty years ago," Zara said, her brow furrowed. "But the supplemental 'infrastructure maintenance' permits? Those were signed off last year. Right before the flow started to drop."

"That's our smoking gun," Declan said, his eyes sharpening with a sudden, dangerous focus. "If we can prove that the maintenance work was actually an unauthorized installation of an industrial-grade pumping system, we don't just have a theory. We have a legal violation."

Micah crossed his arms, his posture shifting from defensive to resolute. "But the people who approved these permits are the same people who sit on the Park Board. If we bring this to them, they'll bury it."

"Not if we don't give them a chance to hide it," Zara said, her journalistic instincts flaring. "If we present this as a public safety issue—toxic runoff entering the community's water source—they can't ignore it. It stops being a park issue and starts being a health crisis."

Avery looked at the others. For the first time, she didn't feel like the girl who just watched the water. She felt like a part of the stream itself—turbulent, necessary, and ready to carve a new path. "We don't need their permission to tell the truth. We need the evidence to be undeniable."

She took the 1974 map and slid it under the tablet. "Declan, I need you to map the exact route of that conduit. If we can show how the suction travels from the Enclave to the spring's heart, it becomes visual. People can't argue with a map."

"I can do that," Declan said, his hands moving across the screen. "I've got the drone footage. I'll overlay the sonar data. It'll be a 3D model of the whole tragedy."

"And I'll dig into the permit history," Zara added, already typing on her laptop. "I'm going to find out who signed those maintenance orders and if there's a paper trail linking them to the development company."

Micah set the cloudy vial down on the desk, his jaw set. "I'll document the ecological damage. I'll map the vegetation shifts and the toxicity levels against the timeline of the pump's installation. If we can show that the water quality dropped exactly when the pumping ramped up, the data will speak for itself."

Avery looked at her friends. The shed was still a wreck, and outside, the spring was still suffering, but the atmosphere had shifted. The panic of the morning had condensed into a cold, hard resolve.

"Okay," Avery said, her voice steady. "We spend the night here. We compile every photo, every sensor reading, and every public record. Tomorrow morning, we don't go to the park office to ask questions. We go to the town hall meeting, and we present the autopsy of our spring."

"And if they try to shut us down?" Micah asked, though the doubt was already leaving his eyes.

Avery looked out the window again, catching the moonlight on the water. It was faint, barely a shimmer, but it was there. "Then we make sure the rest of the town is watching."

They worked in a rhythm born of necessity. For the next several hours, the shed transformed into a war room. Declan's drone footage played on a loop, the murky depths of the conduit system revealed in haunting grayscale. Zara's research began to fill the whiteboard, connecting names from the planning board to the developers of the Enclave with red string and frantic marker strokes. Micah painstakingly labeled water samples, creating a timeline of chemical spikes that aligned perfectly with the development's expansion phases.

Avery stood at the center of it all, her role no longer just an observer. She was the one synthesizing the pieces—the geology of the basin, the politics of the permits, and the biology of the dying spring. She felt the weight of it, the responsibility, but it wasn't a burden anymore. It was a purpose.

As the first gray light of dawn began to seep through the dust-streaked windows, they finished the final slide of their presentation. It was a simple, damning story: a map showing how a community had traded its lifeblood for convenience, and how the infrastructure of their own progress had turned on them.

Declan leaned back, his chair creaking. "It's ready."

Avery reached out and touched the screen. The image showed the spring as it was, and the path of the poison that was killing it. "This is it," she whispered. "This is how we give the water back its voice."

They stood up together, the collective exhaustion forgotten. The mystery was solved. The investigation was over. Now, the real work began.

Chapter Seven

The Limestone Archive

The morning air clung to the limestone banks of Glasswater Spring like a damp, heavy wool blanket. The silence, usually punctuated by the rhythmic lap of water against cypress knees, was now jagged and wrong.

Avery stepped onto a shelf of porous, cream-colored rock that, only a month ago, had been three feet underwater. It was slick with dried algae—a brittle, ghostly skin that crackled under her sneakers. She looked back at Declan, who was already hunched over a tripod, his face lit by the cold blue glow of a tablet.

"The pressure differential," Declan muttered, not looking up. "It's not just the drought, Avery. Look at the bedding planes. This limestone is shifting under its own weight. It's losing the hydraulic support it's had for ten thousand years."

Avery knelt, pressing her palm against the stone. It felt hollow, vibrating with a deep, subterranean fatigue. "It feels like it's gasping, Declan. Like the whole spring is trying to hold its breath."

"That's not the spring," Declan said, shifting his drone's landing pad aside. "That's the aquifer collapsing into the voids created by the Enclave's extraction. Every gallon they pull out is a gallon taken from the structural integrity of this entire bank."

Avery walked further along the exposed ledge, her heart sinking. She recognized this spot. This was the "Cathedral," a deep-water pocket where manatees used to congregate. Now, it was a mud-choked depression. A small, stranded garfish flailed in a shrinking puddle, its scales dull in the filtered sunlight. Without thinking, she scooped it up—the fish cool and frantic in her hands—and moved it to the deeper, main channel.

"You can't save them all, Avery," Declan said, his voice softening.

"I know," she whispered. She looked at the garfish darting away into the murky depths. "But I have to know why it's happening. You see the data, but I... I feel the rhythm. The rhythm is broken. It's not just a drop; it's an extraction. It's theft."

Declan walked over, his boots crunching on the desiccated snail shells. He held his probe out toward a fissure in the limestone—a jagged, vertical crack that hadn't been there two days ago. "Look at the tilt. The bedrock is tilting toward the south, right toward where the Enclave installed their high-capacity industrial pump. This isn't just a geological anomaly. It's a direct consequence of the stress they've applied to the conduit."

Avery leaned in. Deep inside the fissure, she saw movement—not water, but a faint, rhythmic pulsing of silt. "It's pulling in the sediment. If the water table drops any further, those conduits won't just dry up. They'll cave in. If they cave in, the spring won't just stop flowing—it'll be choked off permanently."

"That's the hard truth," Declan said, his voice tight. "The science doesn't lie. This is a structural failure in progress."

Avery looked up, her gaze tracing the line of the old cypress canopy. Her intuition, which had always served as a compass, suddenly aligned with

Declan's cold, hard numbers. The "rhythm" she sensed was the sound of the earth settling into the void left by the water.

"It's not just the pump, Declan," Avery said, her voice growing steady. "I remember my grandfather talking about the 'original flow.' He used to say the spring wasn't just a pool; it was a vein. It connected to a system that spanned the entire county. If they've pinched the vein here, they've disrupted the entire flow of the region."

Declan straightened up, his brow furrowed. "A system? You mean the old floodplains?"

"Exactly," Avery said. She pulled a worn, plastic-covered field journal from her pack. She flipped past sketches of birds and water lilies until she reached a rough, hand-drawn map she'd been building since the spring began to fail. "Look. When we combine the drone survey data you captured yesterday with these old survey markers from the 1950s that my grandfather kept... the extraction points they're using align perfectly with the original, natural recharge pathways."

Declan squinted at her notebook, tracing the lines with a gloved finger. "They're not just pumping from the spring. They're intercepting the water *before* it even reaches the recharge zone. They've effectively placed a dam underground."

"If we can prove that these markers are historically documented as vital to the spring's health, the development permits they're hiding behind might not be valid," Avery said, her eyes flashing with a sudden, sharp clarity. "They've bypassed the environmental impact protocols by claiming they were only tapping into 'dead' zones."

Declan looked back at the drying bed, then at his equipment. A surge of adrenaline replaced the pragmatism in his eyes. "That changes everything. It's not just environmental negligence—it's a violation of the land use charters established when the park was founded."

"We need to see the original records," Avery said, closing her journal with a snap. "Not the digital summaries the town clerk keeps on the website. I want the physical ledger, the one from the original survey. If those maps exist, they'll show exactly how the water is supposed to move."

Declan began packing his gear, his movements quick and decisive. "If the archive is as dusty as I think it is, we're going to be digging through ghosts."

"Then we'll dig," Avery replied, casting one last look at the exposed, dying bank of the spring. The silence of the water—the absence of its usual, vibrant song—felt like a physical weight on her chest. But for the first time, she felt she had the tool to fight it. "Let's go find the proof."

They turned away from the limestone, the sun now high and white-hot over the canopy. As they trekked back toward the research shed, the air smelled of ozone and damp earth. They weren't just two students walking through a park anymore; they were mapping the anatomy of a crime. The ground beneath their feet, once a source of comfort, now felt like a mystery waiting to be decoded.

Avery quickened her pace, the weight of the task ahead grounding her. She knew that beyond the park gates, the town was getting ready for the meeting, oblivious to the fact that their heartbeat was being squeezed out of existence. But as she saw Declan check the battery life on his drone, she knew they wouldn't just show up with an opinion. They were going to show up with the maps of the forgotten, and they were going to force the town to look at the truth.

Declan wiped a smudge of red clay from his tablet screen, his brow furrowed as he toggled between the sonar scan and the geological overlay. The limestone shelf beneath them groaned—a low, rhythmic thrumming sound that vibrated through the soles of Avery's boots. It wasn't the sound of the earth settling; it was the sound of a system under artificial tension.

"The pressure differential is too sharp," Declan muttered, more to the rock face than to her. "If this were a natural drought, the drawdown would be gradual, buffered by the saturation of the surrounding sediment. But this? The permeability is being choked off from the inside out."

Avery crouched low, her fingertips tracing the jagged, pale edge of the exposed limestone. Usually, this spot was submerged in three feet of crystal-clear water, teeming with grazing manatees and schools of sunfish. Now, it was a tomb of damp, suffocating moss. She looked at the rock not as a geologist, but as a witness. She remembered the way the water used to swirl here—a lazy, clockwise eddy that suggested a deep, open conduit beneath the surface.

"It's not just drying, Declan," she said, her voice quiet. "It's being stifled. It's like the spring is trying to breathe through a straw that someone is pinching shut."

Declan shifted, his gear clattering softly against his harness. He pointed his laser rangefinder at a cluster of stalactites hanging from a ledge just above the waterline. "Look at the mineral staining. That line there? That's where the water level sat for fifty years. The new line, the one that's currently bone dry, is twenty inches above it. We aren't just seeing a drop in the water table. We're seeing a structural shift in the flow path. The pumping at The Enclave isn't just taking water; it's physically rearranging the subterranean plumbing."

Avery pulled a crumpled, laminated field map from her vest—one she'd snagged from the local historical society weeks ago. It was a hand-drawn survey from 1952, detailing the original hydrology of the Glasswater basin before subdivisions and drainage culverts had sliced up the area.

She spread it over a flat slab of limestone, using a heavy river stone to weigh down the corners. "If your sonar data is right," she said, pointing to a pencil-marked star on the map, "then the primary conduit connects here, under the old citrus grove, right toward the northern development boundary."

Declan leaned over, his eyes scanning the faded ink. "That alignment... It's perfect. It matches the anomaly I picked up on the drone's thermal feed this morning. The temperature variance in that exact spot suggests a massive volume of cool groundwater being diverted into a high-pressure intake pipe."

"They didn't just build on top of the recharge zone," Avery realized, the pieces finally snapping together with a sickening clarity. "They tapped the vein. They turned the heart of the spring into their own private reservoir."

She felt a surge of cold anger, but she forced it back, breathing in the scent of damp earth and decaying leaves. She couldn't afford to be just angry; she had to be surgical. "We have enough to prove *what* they're doing. But if we want to force a stop-work order, we need to prove it's illegal. We need to show that this land was protected under the original zoning, back before they carved out their 'emergency' exemptions."

Declan packed his tablet into his waterproof case, his movements sharp and efficient. "The town hall archives. Zara mentioned the municipal records from the late nineties are still physical. They haven't been digitized."

Avery stood up, her joints aching from the humidity and the tension. She looked out over the shrinking basin one last time. The water was still clear, but it was stagnant, a mirror reflecting a landscape that was rapidly losing its lifeblood. "We're going to need the original surveyors' reports. If we can show that the hydrological impact study for the initial Enclave permit was based on flawed or ignored data, the whole legal foundation for their pumps falls apart."

"And if we find the discrepancy, we take it to the town board," Declan said, his tone turning grim. "No more 'teenage alarmism.' Just the facts, plotted against their own historical neglect."

They climbed up the muddy bank, leaving the silence of the spring behind. As they scrambled up the incline toward the trailhead, the humid air of the Florida morning hit them with renewed force, heavy with the smell of pines and distant asphalt.

"My truck is at the visitor center," Declan said, shielding his eyes from the rising sun. "We can be at the archives by the time they open. Avery—are you sure about this? Once we present this, there's no walking it back. The developers aren't going to apologize and turn off the taps."

Avery looked back at the spring, where the water levels seemed to have dropped another fraction of an inch in the time it took to climb the bank. The reality of the subterranean collapse wasn't just data points on a screen; it was the loss of the only place that had ever felt like home.

"They won't apologize," she agreed, her voice steady. "But they'll have to listen. The land isn't quiet anymore, Declan. It's screaming. We're just the ones translating it."

They reached the gravel parking lot, their boots heavy with the clay of the spring floor. The morning sun glared off the windshield of Declan's truck, but the light felt different to Avery now—not like the start of a Discovery Day, but like the opening of a case file. As they climbed into the cab, the hum of the road ahead felt like a necessary departure from the sanctuary they had just left. They weren't just students anymore; they were the architects of a disruption, and the town of Glasswater was about to realize that the spring was not as silent as they had hoped.

Chapter Eight

Questions Under the Boardwalk

The sun hung low over Glasswater Spring, casting long, bruised shadows through the cypress knees that poked out of the receding waterline like skeletal fingers. The humidity was a physical weight, clinging to the skin, but Avery felt a different kind of chill as she crouched beneath the wooden slats of the main observation deck.

Above them, the familiar, rhythmic thud of tourist footsteps drummed against the planks. Avery peered through the gaps in the timber, focusing on the shoreline. Where the vibrant, emerald-green hydrilla should have been swaying in a vigorous current, there was now only exposed, slick limestone, coated in a thick, dying slime.

Zara was pressed against a support piling, her notebook open, a digital recorder clutched in her other hand. She tilted her head, listening to the strained, practiced cheerfulness of Ranger Miller's voice carrying down from the deck above.

"It's just the seasonal oscillation, folks," Miller was saying to a small knot of confused visitors. "The Floridan Aquifer is a complex system. These lower levels? They're just part of the natural rhythm. We've had a dry spell, the vegetation is taking a nap, but the spring will bounce back. Nothing to worry about."

Avery tightened her grip on her camera. The data they had scraped from the research shed proved Miller was lying. The flow rate wasn't oscillating; it was flatlining.

"He's doing it again," Avery whispered, her voice barely audible over the drone of a nearby cicada. "He's treating it like a PR problem instead of a death spiral."

"He's reciting a script," Zara muttered, her eyes narrowing. She checked her watch. "He's off-duty in ten minutes. If we don't get him now, he'll slip out the back gate, and we'll have to wait until tomorrow."

Zara didn't wait for a reply. She scrambled out from under the boardwalk, smoothing her shirt and adopting a look of polished, wide-eyed curiosity. Avery followed, heart hammering against her ribs.

Ranger Miller was a man whose tan seemed permanently etched into his skin, a testament to twenty years of Florida sun. He looked tired, the kind of exhaustion that had nothing to do with physical labor and everything to do with carrying a secret. He was pinning a "Restricted Access" sign to the railing near the spring's primary outflow.

"Excuse me, Ranger Miller?" Zara's voice was bright, professional, the tone of a student reporter looking for a human-interest story.

Miller flinched, then regained his composure, offering a tight, thin-lipped smile. "Afternoon. You two shouldn't be back here. Park's closing in twenty."

"We're just working on a project for the Spring Discovery Day," Zara said, stepping into his personal space, just enough to be unavoidable. "I was looking at the historical flow logs, and I'm a bit confused. The archives from the nineties show that even during the major drought of '98, the discharge never dipped below fifty cubic feet per second. Today, the gauge at the basin is reading thirty-eight. If this is just a 'seasonal rhythm,' why are the numbers dropping faster than they did in a declared emergency?"

Miller's smile didn't reach his eyes. He tightened the bolt on the sign. "Numbers fluctuate, kid. Sensors get calcified. You need to stop digging into stuff you don't understand."

"I understand basic math," Zara countered, her voice dropping an octave, losing the 'student' affectation. "I also understand that you filed a maintenance report three weeks ago citing 'routine clearing of debris.' But the drone footage I've reviewed shows no clearing. In fact, it shows the construction of a permanent, high-capacity intake manifold buried under the sediment near the northern grotto. Why is the park installing industrial-grade infrastructure while the spring is actively drying up?"

The color drained from Miller's face, leaving him looking sallow. He glanced toward the administration office, then back at the two girls. "You're making assumptions. That 'infrastructure'—as you call it—is for fire suppression. For the park's safety."

"Fire suppression?" Avery spoke up, her voice trembling slightly but clear. She stepped forward, pointing toward the murky, low-flow zone. "I've spent every weekend of my life in this water. I know the current. I know that the spring isn't being suppressed; it's being diverted. I saw the turbidity sensors spiking, Ranger. That's not fire suppression. That's a pump station pulling groundwater before it even reaches the basin."

Miller looked at Avery, really looked at her, and for a second, his defensive wall cracked. He saw the genuine, agonizing hurt in her eyes—the look of someone watching a friend die. He looked down at the boardwalk, his jaw working as he chewed on his lip.

"You kids," he whispered, his voice raspy. "You think you're uncovering a conspiracy, but you're just making yourselves targets. Some decisions are made in offices you don't have access to, by people who don't care about the temperature of the water or the health of the eelgrass."

"Then tell us who," Zara pressed, hitting the 'record' button on her device. "If it's institutional, let's put it on the record. If the state is killing Glasswater, the public has a right to know."

Miller stepped closer, lowering his voice until it was a harsh, conspiratorial rasp. "I don't answer to the public; I answer to the Board of Trustees. And they signed off on that 'maintenance' back in January. The Enclave development needed a 'supplemental water source' to maintain their golf course and the fancy koi ponds. They called it a 'public-private partnership.' I'm just the guy who has to put up the signs."

Avery felt the world tilt. The Enclave—the sprawling, gated luxury development that had been eating away at the surrounding woods for the last two years. She had watched the trees fall, but she hadn't connected the clearing of the land to the silencing of the spring.

"You're siphoning the spring to water grass in a gated community?" Avery asked, her voice tight with disbelief.

Miller let out a humorless laugh. "I'm not doing anything. I'm just trying to keep my pension until retirement. If you want answers, you stop looking at the spring, and you start looking at the land deeds. The water rights don't belong to the spring anymore. They sold the rights to the future."

He stepped past them, his heavy boots thudding hollowly on the wood. He didn't look back.

Avery stood frozen, watching the water ripple listlessly against the limestone. The "seasonal rhythm" excuse was gone, replaced by something much colder and more calculated. This wasn't a natural disaster. It was a theft.

Zara stood beside her, looking down at the digital recorder. Her hands were shaking, but her expression was one of grim determination. "He gave us the thread, Avery. The land deeds. The Board of Trustees."

"It's not just a drought," Avery whispered, feeling the weight of the realization. "It's a managed failure. They're treating our home like an open faucet."

"Then we're going to have to turn the water off at the source," Zara said, tapping her notebook against her palm. "Come on. We need to get back to the shed. If we're going to fight the Board, we need more than just flow rates. We need to prove that they knew exactly what they were doing when they signed those permits."

Avery took one last look at the spring. The water seemed to darken, the low level revealing jagged, broken pieces of rock that had been hidden for decades. She wasn't just an observer anymore; she was a witness to a crime. As they walked toward the exit, leaving the park behind, she felt the silence of the spring following them—not a peaceful quiet, but a waiting, desperate stillness that demanded they come back with the truth.

The humidity under the boardwalk was a physical weight, thick with the smell of rotting hydrilla and stagnant, sulfurous mud. Avery wiped a smear of damp silt from her forehead, her fingers trembling—not from the heat, but from the adrenaline of what they had just heard.

Above them, the heavy tread of tourists signaled the shift change at the park's main gate. A few yards away, near the intake grate, Ranger Miller was busy hammering a "Low Water Conditions" sign into the soft, receding bank. He looked frazzled, his uniform wrinkled, avoiding the eyes of a frustrated family trying to launch a paddleboard into six inches of water.

Zara stayed crouched behind a support piling, her digital recorder gripped tight. She nudged Avery, pointing toward the maintenance shed adjacent to the utility access gate. A man in a grease-stained polo shirt—Hank, the lead maintenance technician—was fiddling with a series of pressure valves that shouldn't have been there. They were industrial-grade, far too robust for the park's small filtration system.

"He's the one," Zara whispered, her voice barely audible over the hum of a nearby cooling fan. "Miller is the mouthpiece, but Hank is the one who turns the screws. If we don't get a direct answer now, we're just guessing."

Avery looked at the spring. The water, usually a vibrant, crystal-clear sapphire, looked bruised. The turquoise depths she had dived into for years were turning a dull, muddy brown. The rhythmic, mechanical *thrum* that Declan had tracked—the sound of the pumps—was audible even here, vibrating through the wooden pilings and into the soles of her sneakers.

"Wait," Avery whispered, grabbing Zara's sleeve. "Look at the intake valve. It's not connected to the spring's circulation. It's pulling from the aquifer directly below the Grotto."

Before Zara could protest, Avery stepped out from the shadows of the boardwalk, her boots crunching on the drying limestone.

Hank jumped, his wrench clattering against the metal housing of the pump. He spun around, eyes widening when he saw the two girls. "What are you doing back here? This is a restricted zone. You need to clear out, now."

"We're not leaving, Hank," Zara said, her voice dropping into her 'journalist' register—steady, calm, and utterly immovable. She didn't hold the recorder up aggressively; she held it naturally, as if it were an extension of her hand. "We know the park's flow rate has dropped by forty percent in three weeks. And we know it's not because of a dry summer."

Hank laughed, a dry, nervous sound. "Look, kid, climate cycles happen. The aquifer is a living thing. Sometimes it breathes, sometimes it doesn't."

"The aquifer doesn't breathe in rhythmic three-minute intervals," Avery interjected, her voice steadying. She pointed to the pump, which had just cycled to a higher pitch. "That's a variable frequency drive. It's calibrated to keep the pressure consistent for the new irrigation lines at the Enclave development. You're not just managing the park's water; you're selling the spring's lifeblood to a gated community."

Hank's face paled. He looked toward the main gate, checking for Ranger Miller, but the ranger was still distracted by the tourists. "You have no idea what you're talking about. I'm just doing maintenance. I get a work order, I do the job. That's all."

"Who signed the order?" Zara stepped closer, invading his personal space just enough to cut off his escape route. "We checked the county records, Hank. There's no public permit for an industrial extraction point on this section of the property. That means this entire setup is illegal. If the environmental protection agency finds out you've been bypassing the flow-control regulators, it's not just the Park Board that's in trouble. It's you."

Hank looked at the pump, then back at the girls. The defiance in his eyes was crumbling, replaced by the crushing weight of a man caught between his paycheck and his conscience. He leaned against the metal piping, the vibration of the water—*the stolen water*—thrumming through his spine.

"It's not just the Enclave," he muttered, his voice cracking. "They pushed through a 'maintenance upgrade' last month. The Board said it was to ensure the springs didn't stagnate if the drought got worse. But the moment those new pumps went online, the pressure in the aquifer dropped. I told them. I told Miller that the sinkhole connection would collapse if we drew more than five thousand gallons an hour. He told me it was 'non-negotiable'—that the development project was too important to the district's budget."

Avery felt a cold spike of dread in her chest. "They knew? They knew it would collapse the sinkholes?"

"They knew it would drain the basin," Hank corrected, looking down at his grease-stained hands. "They just didn't care. They thought the 'natural cycle' would cover the loss."

Zara didn't waste a second. "Hank, we need that work order. We need the original, signed, stamped document from the Board. We have a meeting in two days, and we're going to present this. If you give us the documentation, you're a whistleblower. If you stay silent and we blow the lid off this, you're just the man who held the wrench."

Hank hesitated. He looked at the spring, his gaze lingering on the exposed roots of the cypress trees that were now clawing at dry air. He reached into his back pocket, pulling out a crumpled maintenance log.

"I kept copies," he whispered, pressing the folder into Zara's hands. "For my own protection. Don't tell them it came from me. If they find out, I'm fired by noon."

"We won't have to," Avery said, her voice filled with a sudden, sharp clarity. "Because by the time we're done, the Board won't have the authority to fire anyone."

As they retreated from the shed and back under the cover of the boardwalk, the sun began to dip below the tree line, casting long, fractured shadows across the dry, cracking sand of the spring's edge. The mechanical hum of the pump remained, a heartbeat that didn't belong to the land, but as Avery looked at the document in Zara's hands, she realized the rhythm was about to be broken.

They walked in silence for a few hundred yards, the weight of the conspiracy finally taking its proper form. It wasn't just a drought. It was a heist.

"We have them," Zara said, her voice shaking slightly now that they were out of earshot. "With these logs, the Board can't just dismiss us as kids playing with science kits. This is a paper trail that leads straight to the pocket of every council member who voted for the Enclave expansion."

Avery looked back at the spring. A small turtle, stranded in a shrinking puddle, scrambled to find deeper water. She felt the urge to go down there, to push the turtle toward the main vent, but she forced herself to stay on the path. They had a bigger goal now. Saving the turtle wasn't enough; they had to save the system that allowed the turtle—and the spring—to exist.

"Declan and Micah are going to lose their minds when they see this," Avery said.

"They're not going just to lose their minds," Zara replied, pulling out her phone to send a coded text to the others. "They're going to help us tear this whole thing down."

The air felt slightly less heavy as they reached the edge of the parking lot. The humidity was still there, but it felt different—charged with the promise of a fight. Avery stopped at the edge of the boardwalk, turning back one last time to watch the water. For the first time, she didn't feel like a passive observer. She was a witness, an investigator, and a protector. And she knew exactly what had to come next.

Chapter Nine

Maps of the Forgotten

The air inside the Lake County Historical Society archives was thick, not with the scent of old paper, but with the cloying, metallic tang of an aging HVAC system struggling against a Florida July. Outside, the heat blurred the edges of the parking lot; inside, the silence was absolute, broken only by the rhythmic *thump-slide* of microfiche reels and the persistent tap of Avery's pen against a legal pad.

Zara pushed a stack of crumbling, salt-stained blueprints across the folding table. "Look at the topography, Avery. Not the roads, not the property lines. The elevation markers."

Avery leaned in, her eyes tracing the topographical lines. She knew the land by its veins—the way water moved through the limestone—but seeing it rendered in the ink of a 1922 surveyor was like looking at an X-ray of a ghost. "These blue shaded areas," Avery murmured, her finger hovering over a vast, irregularly shaped patch that now sat beneath the sprawling concrete footprint of the modern shopping plaza and the outskirts of The Enclave. "These aren't just low spots. They're recharge zones."

"Exactly," Zara said, her voice dropping to a conspiratorial whisper. "Those aren't just marshes. They were the lungs of the spring. The rain hit these depressions, sat in the sawgrass, and percolated down into the Floridan

Aquifer. It kept the pressure constant. Now? It's all storm drains, retention ponds, and asphalt. The water doesn't soak in; it gets diverted into the canal system."

Avery pulled a current municipal map from her bag, layering it over the brittle historical document. The misalignment was jarring. "They built the parking lots and the new housing developments right on top of the primary intake valves for the basin. No wonder the spring is choking. It's not just the pumps at The Enclave; it's the entire landscape. We've paved over its ability to breathe."

"It's not just an oversight," Zara noted, pulling a heavy, leather-bound property ledger toward her. "Look at the dates on the rezoning permits. 1984. 1996. 2008. Every time a new developer came in, they requested 'land surface modifications' to prevent flooding. They turned natural wetlands into engineered drainage basins to save their foundations from water damage. They didn't just ignore the spring—they actively blocked its source to dry out the land for property values."

Avery felt a chill that had nothing to do with the struggling AC. "If they knew—if the planning board had these maps—they knew the recharge zones were vital."

"They knew," Zara confirmed, her jaw tightening. "Which means this isn't just about a few 'emergency' pumps. This is a forty-year history of prioritizing real estate over the aquifer. We're not fighting a drought, Avery. We're fighting a legacy."

The heavy oak door at the front of the archive creaked open, admitting a slice of blinding, humid sunlight. Mr. Henderson, the town clerk, shuffled in, his spectacles perched precariously on the end of his nose. He was a man who measured his life in administrative filings and kept the town's secrets buried in gray steel cabinets.

"Closing time," he grunted, not looking up from his ledger. "Society's closed to the public on Tuesdays after four."

Zara didn't pull the documents away. Instead, she slid the 1922 map directly into his line of sight. "Mr. Henderson, I'm working on a story about the aquifer's decline. I think you'll find these records... enlightening."

Henderson glanced down, his indifference faltering as he recognized the old district seal. He adjusted his glasses, his eyes widening as they hit the shaded blue zones. "Where did you get these? These were archived for the state assessment. They aren't for general distribution."

"They're public record, aren't they?" Zara pressed, her voice sharp with journalistic steel. "Or is the public not supposed to know that the town's most affluent developments were built directly on top of the critical recharge zones for Glasswater?"

Henderson stiffened, his fingers trembling slightly as he reached for the map. "You're looking for a scandal where there's only progress. That land was swampy, unusable. Development brought jobs. It brought the tax base that keeps this town afloat."

"It brought a dry spring," Avery interrupted, her voice quiet but steady. She stood, her presence surprisingly formidable in the cramped room. "The flow isn't just low, Mr. Henderson. It's dying. The vegetation in the grotto is changing. If we don't restore the permeability of those zones, the spring won't just struggle—it will disappear."

"The town board isn't going to tear up the Enclave's foundation because a couple of kids are worried about water levels," Henderson retorted, his face reddening. "That's property owner equity. You go stirring that pot, you're looking at lawsuits that'll bankrupt the town historical society before you even get a headline."

"Then explain this," Zara said, pulling the final, most damning document from her satchel. It was a faded, handwritten amendment to an irrigation permit from the late seventies—an anomaly that had eluded even the most thorough digital searches. "What's this? An 'Inter-basin transfer' permit granted to the original grove owners, grandfathered in for 'essential agricultural maintenance'?"

Henderson looked at the paper, then back at the two girls. His face drained of color. "That permit was for the citrus trees. It shouldn't be active. The groves were cleared thirty years ago."

"But it is active," Avery said, recalling the readings Declan had pulled from the pump sensors. "The pump at the Enclave is pulling from that old wellhead. They didn't get a new permit. They're using a defunct agricultural permit to bypass the current environmental regulations, aren't they?"

Henderson looked at the door, then back at the map, his loyalty to the board clearly warring with the sheer, undeniable weight of the evidence on the table. "You girls have no idea what you're doing. You're talking about the families who built this town. Their money, their influence—it runs the water board, the planning committee, and half the local businesses."

"Good," Zara said, clicking her pen with a finality that echoed off the rows of dusty shelves. "Then they'll be able to afford the restoration fees."

Henderson let out a shaky sigh, his shoulders slumping. He reached into his vest pocket, pulled out a key, and unlocked a small, nondescript cabinet behind his desk. He pulled out a thick, dust-coated folder and tossed it onto the table. "I didn't give you this. I don't know what's in it. But if you're going to drag the board into the light, you'd better make sure you have every single page. Because they won't just fight you in the press—they'll fight you in the court."

Avery opened the folder. Inside were the blueprints for the original irrigation infrastructure, complete with handwritten notes in the margins—notes that laid out, in clear, cold detail, exactly how the drainage patterns were designed to divert water away from the spring's basin to favor the developments.

"They didn't just neglect the spring," Avery whispered, her finger tracing a red ink annotation on the print. "They engineered its demise."

"We have everything now," Zara said, her eyes flashing with a mix of triumph and righteous anger. "Every permit, every blueprint, every lie signed by someone who knew exactly what they were stealing."

Avery looked at the map one last time, seeing not just lines and colors, but the history of her home. The spring wasn't just a place to kayak; it was a living, breathing system that had been betrayed by the very people who claimed to cherish it.

"Let's go," Avery said, gathering the documents. "Micah and Declan are waiting. It's time to show them what's actually flowing through their taps."

The hum of the industrial air conditioner in the town clerk's office was a rhythmic, dying wheeze, struggling against the humid pressure of a mid-July afternoon. Zara dropped a heavy, plastic-sleeved document onto the desk with a sharp *thwack*. The sound echoed off the cramped walls, making the Town Clerk, Mr. Henderson, flinch behind his stack of tax assessments.

Avery stood back, her eyes tracing the yellowed edges of the 1928 surveyor's map spread out like a shroud across the counter. She felt the weight of the past in the room—the way the paper crinkled, the smell of damp earth and slow-decaying cellulose.

"Mr. Henderson," Zara said, her voice steady, stripped of the conversational lilt she usually kept for her blog. "We've spent the last six hours reconciling these property deeds with the 1928 hydrological survey. It's not just a coincidence that the flow rate dropped the moment The Enclave finished their north-ridge expansion."

Henderson adjusted his thick-rimmed glasses, not looking at the map. "You two are wandering into deep water. Planning decisions were made by people who've been in this town longer than you've been breathing.

Development brings tax revenue. Revenue brings roads, schools, and services."

"It brings a dead spring, too," Avery added, her voice quiet but piercing. She pointed a finger at a sprawling, shaded-in area on the antique map. "This wasn't just low-lying ground. This was the Cypress Basin. It was a primary recharge zone for the entire limestone shelf. When the developers paved over this to build the Enclave's main access road, they didn't just pour concrete—they sealed the lungs of the aquifer."

Henderson scoffed, though his hand trembled slightly as he reached for a lukewarm cup of coffee. "You're reading too much into lines on a page. That land was swampy, unusable. It was a net positive to stabilize it."

"Stabilize it by siphoning water straight from the under-bed with 1970s agricultural irrigation permits?" Zara leaned in, her phone recording, the red light blinking like an accusing eye. "We found the permit, Mr. Henderson. File 88-Beta. It's a grandfathered agricultural extraction permit meant for citrus groves that haven't existed in forty years. The Enclave is using it to water their manicured lawns and fill their decorative ponds, while the state park's primary flow is being starved to compensate for the pressure drop."

The door to the office creaked open. An older man, Mr. Abernathy—a local fixture known for his stubborn devotion to the town's founding families—stepped in, hat in hand. He stopped when he saw the girls, his eyes narrowing as he recognized the map on the desk.

"You're making a mess of things, girls," Abernathy said, his voice raspy. "That land you're pointing at? My grandfather sold that parcel. It was an investment. You want to tear up the road? You want to bankrupt the people who keep this town afloat?"

Avery stepped forward, her movements fluid, guided by the same instinct she used to navigate the spring's hidden channels. She didn't look at the adults with fear; she looked at them with the clarity of someone who had spent her life watching water run where it was meant to go.

"Mr. Abernathy, I don't want to bankrupt anyone," Avery said, her voice dropping into that calm, resonant tone that Micah always said could de-escalate a storm. "But you remember the flood of '94, don't you? Before they built the diversion canals and the paving? The spring surged, it cleared the silt, and the basin flourished. It's a cycle. You've broken the cycle. You didn't just buy land; you bought the water rights to our future, and you're wasting them on Bermuda grass."

"It's not just waste," Zara added, seizing the moment of silence. "It's fraud. You're using agricultural tax exemptions for luxury real estate. I have the receipts, I have the maps, and I have the names of the board members who signed off on the 're-zoning' of a recharge zone without an updated environmental impact study."

Abernathy's face flushed a deep, mottled red. "You think you're so smart. You think a few college applications and a digital camera make you the conscience of the county. You have no idea how hard it is to keep a community from sliding into poverty."

"I know the price of the spring," Avery replied, pointing at the map again, specifically at a faded red ink marking that denoted a sinkhole conduit. "And if that goes dry, the sinkholes in the residential sector will collapse, because the pressure from the aquifer is the only thing keeping the roof of those caves from falling in. You're not just killing the park. You're undermining the foundation of every house in the Enclave."

Henderson went pale. The math was simple, and he knew it. If the water table dropped below a certain threshold, the geological structural integrity of the entire North Ridge would vanish.

"That's just a theory," Henderson stammered, his bravado crumbling.

"It's a geological certainty," Zara countered, her gaze unwavering. "And I've already sent the digital copies of these blueprints and the permit logs to the Regional Water Management District. We aren't here to ask for your permission to stop it. We're here to give you the chance to fix it before the subpoena hits your desk."

The room fell into an oppressive silence, punctuated only by the clicking of the ceiling fan. Abernathy looked at the map, then at Avery, then at the heavy binder of evidence Zara held like a shield. The defiance in the older man's eyes flickered, replaced by a weary, hollow realization. He knew the girls weren't bluffing; they had the one thing the town's power brokers had lacked for thirty years: the truth, mapped out in ink and iron-clad data.

"You've stirred up a hornet's nest," Abernathy muttered, turning toward the door. "Those families—they won't thank you for this. They'll come after you."

"Let them," Zara said, pulling the map toward her and rolling it up. "The spring doesn't negotiate with the people who block its path. Neither do we."

As they walked out into the blinding Florida heat, the air felt different—thinner, yet charged with a static intensity. Avery took a deep breath, the humidity sticking to her skin, but for the first time in weeks, her chest didn't feel tight. She felt like a current finally breaking through a bottleneck.

"We have them," Zara whispered, checking her phone. "The District office confirmed receipt of the files. They're already deploying a monitoring team to the north ridge."

Avery looked toward the distant tree line where the Glasswater Spring lay hidden in its mossy, drying basin. "This is just the start, Zara. Getting them to stop pumping is one thing. Getting the land to remember how to drink again... that's going to be the hard part."

"One step at a time," Zara said, though her hands were shaking. "The Town Hall meeting tonight? It's not going to be a debate anymore. It's going to be a reckoning."

Avery nodded, her eyes fixed on the horizon. She could hear it—not the sound of the spring, not yet—but the distant, low rumble of change. The town was finally going to have to look beneath the surface.

Chapter Ten

The First Flow Analysis

The surface of the Glasswater Vent was a mirror of dark, unmoving glass. It was wrong. Even at its lowest seasonal ebb, the spring always breathed—a subtle, rhythmic pulse of clarity that kept the silt from settling. Today, it was stagnant.

Avery knelt on the limestone shelf, her GoPro rig strapped to a weighted frame. She didn't look at the water; she looked at the *depth*. Beside her, Declan was hunched over his tablet, the screen shielded by a makeshift cardboard hood.

"The sensor's calibrated to the thousandth, but it doesn't matter," Declan muttered, his brow furrowed. "The telemetry from the park's flow-meter at the pump house says they're pulling a steady four hundred gallons per minute. That's their 'sustainable limit' for drought mitigation."

Avery submerged the probe. She watched the readout in real-time, her heart sinking as the numbers flickered. "Declan, look at the discharge rate at the mouth of the run. If they were pulling four hundred, the vent velocity should still be pushing at least fifty percent of its baseline. We're at eight percent."

Declan squinted at her screen, then tapped his own furiously, cross-referencing the park's public-facing dashboard. "That's physically impossible. If the extraction rate is accurate, the hydraulic head pressure shouldn't have dropped this sharply. It's like the water is being bypassed before it even reaches the headspring."

"Or they're lying about the extraction points," Avery said, her voice tight. She pulled the probe up, droplets of water clinging to the wire like liquid jewels. "They've been telling the public the pumps are pulling from the peripheral aquifer to save the main basin. But if the main vent is dying while the pumps are 'at limit,' the suction isn't peripheral. It's localized. They're tapping the vein, Declan. Directly into the throat of the spring."

Declan stood up, his boots scraping against the limestone. He walked to the edge of the water, looking down into the crystalline abyss. The cavern entrance—the beautiful, dangerous throat of the spring—looked like a dried-up well. "If they've dropped the intake lines directly into the primary conduit, they're not just managing flow. They're cannibalizing the source. The park board's official projections are based on the assumption of a distributed drawdown. If this is centralized, the whole structural integrity of the limestone shelf is at risk. It's not just a drop in water levels, Avery. It's a collapse waiting to happen."

Avery looked toward the distant, manicured lawns of the "Enclave" development visible through the treeline. The irony was suffocating. "They're greening their golf course with the lifeblood of the basin."

"We need more than just flow rates," Declan said, his voice hardening into a geographer's resolve. "We need to show the exact delta between their reported output and the physical reality of the vent. If I can overlay the sonar map I pulled last night with the location of those industrial pumps, I can prove the siphon point is illegal under current state water management codes."

Avery stood, wiping her hands on her jeans. The quiet observer in her was gone, replaced by a cold, sharp anger. She looked at the vent one last time, promising it a fight. "The Regional Water Management District won't ignore hard data, Declan. They're bureaucrats—they speak in numbers. Let's give them a language they can't dismiss."

"We have to be fast," Declan said, already packing his equipment. "If they realize we've found the variance between their 'official' numbers and the ground truth, they'll purge the logs before we get to the meeting."

Avery nodded. She didn't need to say anything else. They didn't move with the casual pace of teenagers on a summer afternoon; they moved with the synchronized urgency of researchers who had just discovered the foundation of their world was being hollowed out beneath their feet.

They scrambled up the embankment, the humid Florida heat pressing down on them, but neither of them felt it. The shock of the discrepancy—the realization that the "mitigation" was actually the sabotage—had sharpened their focus. As they reached the tree line, Avery checked her watch. They were behind schedule, and the town hall meeting felt like a closing door.

"To the shed," she said, her stride lengthening.

"To the shed," Declan echoed, his eyes fixed on the path ahead, already running the calculations in his head. They were no longer just watching the spring; they were preparing to defend it.

The research shed smelled of damp paper, ozone, and the metallic tang of Declan's overworked laptop charger. Outside, the Florida night hummed—a chorus of cicadas and tree frogs that felt cruelly indifferent to the tragedy unfolding beneath the limestone shelf.

Avery pulled a stack of creased, yellowing maps across the scarred wooden table, clearing a space between a box of water-testing vials and a half-empty bag of pretzels. She traced a finger over the red ink markings she and Zara had made earlier.

"Look," Avery said, her voice steady despite the adrenaline still jittering in her chest. "The Enclave isn't just pulling water from the aquifer. They're effectively creating a localized cone of depression. If we look at the recharge rates from the last decade, we've been operating on a deficit since the first foundation for those luxury estates was poured. The math doesn't just suggest a problem; it screams it."

Declan, eyes bloodshot from staring at the sonar mapping software, pushed his glasses up his nose. He tapped a key, pulling up a 3D visualization of the karst conduit he'd been tracking. "You're right, Aves. The pressure differential is catastrophic. I've been mapping the flow resistance along the primary vent, and the data matches the drop-off we saw this morning. It's not just a surface-level diversion—it's a systemic siphon. They're literally pulling the 'breath' out of the spring."

Zara leaned in, her notepad overflowing with scrawled transcriptions from the interviews she'd conducted with the long-term residents. "And here's the kicker," she said, tapping a section of the legal archive. "I cross-referenced the current irrigation permits with the historical land deeds. The park's 'official' records conveniently skip over the 2014 zoning amendment. They reclassified the land usage to 'commercial-agricultural maintenance,' which basically allowed the developers to bypass the standard water-usage caps. It was a bureaucratic sleight of hand. They weren't just careless; they were precise."

Micah, who had been sitting quietly by the window, watching the moonlight filter through the dense canopy of the cypress trees, finally turned back to the group. He held a sample vial up to the dim overhead light. The water inside, usually crystal clear, carried a faint, murky suspension.

"It's not just about the volume of water lost," Micah added, his tone somber. "It's the composition. I've been looking at the nutrient levels in the last twenty-four hours. We're seeing a massive spike in nitrates. It's not just the drought; it's the return flow. The Enclave's landscaping—all that fertilizer—it's leaking directly into the groundwater. It's creating an algal bloom potential that could suffocate the remaining eelgrass in weeks, not years. If the spring stops flowing, the filtration stops, and the spring becomes a stagnant pond."

The weight of their findings hung in the air, heavier than the humidity pressing against the screens. For a long moment, nobody spoke. The silence wasn't the comfortable, companionable quiet they had started with; it was the heavy, clinical silence of a diagnosis.

Avery looked at the map, then at Declan's 3D model, then at Zara's stack of incriminating evidence. Individually, they were just bits of information—a quirky hobby, a school project, a curious news angle. Together, they were an indictment.

"We have to stop looking at these as separate issues," Avery said, her gaze firm as she looked at each of them in turn. "The geology, the permits, the ecology—it's all one chain reaction. If we go to the Town Hall meeting and talk about 'declining water levels,' they'll bury us in 'drought management' rhetoric. They'll blame the climate. They'll blame the heat."

"She's right," Declan said, his adventurous, impulsive energy replaced by a cold, sharp focus. "We need to show the causality. We need to walk them through the physical mechanism of the theft."

"And I'll provide the context," Zara jumped in, her pen poised. "I'll frame it as a breach of public trust. The people of this town deserve to know that the park board—the very people tasked with protecting Glasswater—effectively handed the keys to a private developer for a lower water bill."

Micah stood up and walked to the table. He didn't have a spreadsheet or a drone, but he had the visceral evidence of what was dying. "And I'll speak to the life of the spring. The visitors come here for the wonder of it, for the connection to something wild. If we lose the spring, we lose the community's identity. That's the emotional hook that makes the data real."

Avery felt a surge of pride, not just in the work they had done, but in the way their disparate perspectives were beginning to interlock like gears in a clock. She took a marker and started drawing lines on the map—connecting the Enclave's pump stations to the limestone conduits Declan had identified, and then drawing the flow of contaminants Micah had measured.

"We need a roadmap," Avery said, sketching a flow chart that linked the history, the science, and the biology into a single narrative. "A timeline of choices. We start with the 2014 zoning shift, show how that led to the installation of the high-capacity pumps in 2016, and then overlay the decline in spring flow from 2017 to today. It's not a story about a dry summer. It's a story about a slow-motion collapse that's been manufactured in plain sight."

Declan started dragging files into a single, master directory on his laptop. "I can render the 3D model to show the drawdown effect. It'll be a visual simulation—something they can't just argue away as 'theoretical.'"

"I'll write the summary," Zara said, her fingers flying across her keyboard. "Hard facts, cited sources, no fluff. We make it so professional, so indisputable, that if they ignore it, they aren't just being negligent—they're being complicit."

Micah picked up the sample vial again, peering through the glass. "And I'll prepare the visual aid for the ecological impact. I have the growth charts for the native plants. If we show them what it looks like now versus what it will look like in six months, we won't need to argue. The reality of it will do the work for us."

Avery looked at the clock on the wall. It was nearly 3:00 AM. The dawn was coming, and with it, the day of the Town Hall. The fear that had been nagging at the edges of her mind—the doubt that four teenagers could ever stand up to a board of directors and a wealthy development company—began to ebb, replaced by a cold, crystalline resolve.

She looked at her team. Declan, who had spent his life looking down at the earth, now understood it had a pulse. Zara, who had always looked at the news from the outside, was now writing the story from the inside. And Micah, the quiet defender, was ready to speak for the things that couldn't ask for help themselves.

"We have everything we need," Avery said, her voice barely a whisper, yet clear enough to cut through the hum of the night. "We've got the past, we've got the present, and we have the proof of what happens if we stay silent."

She closed the folder of maps, the snap of the plastic tab sounding like a gavel in the small, crowded room.

"Let's get to work."

Chapter Eleven

ECHOES IN THE GROTTO

The air inside the entrance to the Grotto of the Weeping Stone was thick, heavy with the scent of damp limestone and long-dead moss.

Avery clicked her headlamp to its highest setting. The beam cut through the gloom, revealing the usual waterline stained into the rock face—a dark, permanent scar of where the surface should be. Today, however, the water sat a good four feet below that mark.

"It's not just lower, Declan," Avery whispered. Her voice sounded thin, swallowed by the cavern walls. "It's gone quiet. Can you hear that?"

Declan, chest-deep in the cooling, tea-colored water, paused. He reached out to stabilize himself against a jagged shelf of karst. He tilted his head, his brow furrowed behind his diving mask, which hung around his neck.

"I don't hear the venturi effect," he murmured, his voice tight. "At this level, the flow should be surging through that limestone neck. It should sound like a freight train. Instead..."

"Instead, it's a graveyard," Avery finished.

They waded deeper. The limestone walls closed in, shifting from a wide, arched chamber to a claustrophobic, tunnel-like passage. Every splash of their boots against the submerged rocks sounded unnaturally loud, amplified by the cavern's acoustics. Avery felt a shiver that had nothing to do with the cool spring water. She had spent her entire life tracking the rhythms of Glasswater. She knew the spring's heartbeat—the rhythmic, pulsing push of water that surged from the aquifer, breathing life into the river.

Now, the rhythm was gone. It was just a slow, sickly dribble.

Declan pulled a waterproof tablet from his dry-bag. He tapped the screen, projecting a 3D wireframe of the grotto's interior. "Look at the sonar pings," he said, holding the light steady. "The main conduit—the primary artery—is completely occluded by something on the other side of that limestone curtain. If the pressure isn't building here, it means the entire system has been bypassed."

"Bypassed by what?" Avery asked, scanning the ceiling. Stalactites hung like skeletal fingers, dripping slowly. *Plip. Plip. Plip.*

"Man-made pressure," Declan said, his voice hardening. "Nature doesn't just stop a flow like this without a major seismic event or a sinkhole collapse, and the readings don't show either. Someone has diverted the primary recharge flow. They've effectively put a tourniquet on the spring's throat."

He moved toward a narrow gap in the rock, where a rusted, calcified grate had been partially pried away—the work of decades of neglect and shifting earth. Avery followed, her camera clutched in her hand, though she found herself hesitating to take photos. This felt less like discovery and more like an autopsy.

As they squeezed through the gap into the upper Grotto, the air quality shifted. It grew stale, tainted by the faint, acrid smell of fertilizer—that unmistakable, cloying sweetness that Micah had spent weeks mapping.

"Avery, look," Declan whispered, pointing his beam toward the floor of the grotto.

The water here wasn't the clear, crystalline blue of the main spring. It was murky, clouded with a thick, grey-white sediment. Floating on the surface, trapped in a swirl of stagnant water against the rock wall, were remnants of vegetation that didn't belong here: non-native ornamental grasses and shredded bits of plastic irrigation tubing.

Avery reached out, her fingers trembling, and plucked a piece of the tubing from the water. "This isn't from the park," she said, her voice barely a breath. "This is from the Enclave construction. It's runoff. The sediment from their clearing, the fertilizer from their landscaping... It's all being pushed into the very recharge zones that feed us."

"They aren't just siphoning water," Declan realized, his eyes wide in the harsh LED light. "They're choking the conduits with debris. The sediment is settling in the limestone fractures, hardening like concrete. If we don't clear this, or if they keep pumping, the blockage becomes permanent. The spring isn't just drying up; it's being buried alive from the inside out."

Avery looked at the dark, silent tunnel ahead. The weight of the realization hit her in the center of her chest. This wasn't a mystery of climate or a quirk of the Florida cycle. It was an engineering crime, executed in the dark, hidden beneath feet of solid rock.

"We have to map this," she said, her quiet voice turning steel-hard. "Every meter. Every blockage. If we can prove that the Enclave's drainage is physically causing the structural collapse of the Grotto, it's not just an environmental issue anymore. It's liability."

Declan nodded, his pragmatic edge returning, though his hand shook slightly as he adjusted his equipment. "I'll start the high-res mapping. If we sync this with the surface drainage maps Zara dug up, we have a map of their guilt."

Avery waded further into the shadows, the silence of the spring pressing against her ears. She finally lifted her camera, the shutter click echoing like a gunshot in the cavern. "We're going to show them exactly what they've done to the heart of this place," she whispered to the darkness. "And then we're going to make sure they can't hide it behind a permit ever again."

She stared into the abyss, no longer the quiet observer of the Glasswater, but a witness to its slow murder, and for the first time, she knew exactly how to stop it.

The air inside the Grotto was thick, humid, and tasted of ancient, stagnant limestone. Avery adjusted her headlamp, the beam cutting a sharp, dusty cone through the darkness. Beneath her fins, the water was no longer a rushing, life-giving current. It was a black, oily mirror, sluggish and cold.

"Look at the waterline," Avery whispered, her voice bouncing off the porous, jagged walls. She pointed her light at a calcified ring etched into the rock, nearly three feet above the current surface. "That's where it should be. That's where it *was* last week."

Declan swam alongside her, his movements methodical. He didn't look at the walls; he looked at the floor of the cavern, his sonar unit blinking rhythmically in his palm. "It's not just the level, Avery. Listen."

Avery held her breath. Usually, the Grotto hummed—the low-frequency vibration of millions of gallons of water forcing its way through the Floridan Aquifer's limestone capillaries. It was the rhythmic, steady pulse that had anchored her entire life. Now, there was only the drip. A slow, agonizing *plink... plink... plink* of water falling from a stalactite into the dark basin.

"It's stopped," she said, a cold prickle of dread tracing her spine. "The pulse is gone."

Declan swam to a narrow fissure at the far end of the chamber, where the limestone walls narrowed into a jagged throat. He jammed his waterproof sensor into the gap. "Micah, are you there?"

The comms unit clipped to his shoulder crackled to life, Micah's voice sounding distorted by the damp cave walls. "I'm here. Reading you loud and clear. What are you seeing?"

"We're at the primary intake point," Declan said, his tone clipped, all business. "The flow rate is negligible. It's effectively a stagnant pool. I'm uploading the thermal data now."

A second later, Micah's voice came back, strained and sharp. "Data's coming through. Declan, look at those nitrate spikes. The concentration is massive. It's not just fertilizer; this is concentrated agricultural runoff, probably pushed directly into the conduit."

Avery swam closer to the fissure, peering into the dark, narrow tunnel. She saw the blockage. It wasn't natural. A jumble of synthetic silt, construction debris, and thick, matted vegetation had wedged itself against a series of limestone teeth.

"It's not a drought," Avery whispered, realization dawning on her with the weight of an anchor. "The Enclave didn't just drill. They've been dumping the waste—the soil they cleared for their pump stations, the runoff from their manicured lawns—it's being funneled right into the secondary aquifer. They're literally choking the heart out of the spring."

"Micah," Declan said, his voice dropping an octave. "The pressure differential here... if this blockage stays, the limestone isn't going just to stop flowing. The pressure build-up behind this debris is going to crack the substrate. We're talking about a localized collapse."

"I see it," Micah's voice replied, quieter now, laced with a terrifying calm. "If the roof of that conduit gives way, the entire filtration system for the local water table is compromised. The spring won't just dry up, Avery. It will become a sinkhole for every pollutant the developers have been pushing into the ground. It'll poison the town's supply."

Avery reached out, her gloved hand trembling as she touched the jagged edge of the limestone. The rock felt brittle, wrong. She thought of the serene, turquoise water where she had spent her childhood, now reduced to this suffocating, poisoned tomb. The mystery had shifted. It wasn't a puzzle to be solved for the sake of curiosity anymore. It was an emergency.

"We have to get out of here," Avery said, her voice echoing too loudly in the narrow space. "If that pressure builds and the walls shift, we're trapped."

"Wait," Declan said, not moving. He held his camera up, framing the debris. "I need the shots. If we don't have visual proof of the blockage—the synthetic sediment alongside the natural limestone—the Board will call it a 'natural geological event.' We need them to see the shovel marks in the clay, the plastic piping fragments."

"Declan, it's not safe!" Avery argued, though she didn't pull away. She saw what he saw: a cigarette butt lodged in the mud, a scrap of PVC pipe caught in a root system that shouldn't have been in a deep-water cave. It was a smoking gun.

"Just ten more seconds," Declan muttered, moving the camera in a slow, sweeping arc. "Micah, can you cross-reference the chemical composition of that runoff with the logs Zara pulled? We need to match the signature of the 'lawn treatment' products from the Enclave's maintenance contracts."

"Doing it now," Micah replied, the sound of keyboard clicking audible through the link. "Wait—Avery, look behind the blockage. The infrared scan shows a secondary, smaller conduit opening about five feet to the left. The water isn't moving, but the temperature is surging. Something is pumping high-pressure heat into that chamber."

Avery swung her light. There, behind a wall of twisted debris, she saw the faint, shimmering distortion of water heated to an unnatural temperature.

It was the discharge from the Enclave's HVAC systems—pumping warm, treated waste directly into the Grotto's hidden lung.

"They're using the cave as a radiator," Avery said, horrified. "They're heating the aquifer. That's why the vegetation is dying. It's not just the nitrates; it's the thermal shock."

"That explains the erratic behavior of the fish," Micah noted. "Everything in the spring is keyed to a narrow temperature range. They've turned a natural sanctuary into a chemical-soaked boiler."

A deep, low groan resonated through the cave—a sound of shifting rock that rattled Avery's teeth. The water around them swirled, a sudden, violent displacement of air bubbles rising from the depths.

"The structure is failing!" Declan yelled, snapping the last photo and grabbing his gear. "Move! Now!"

They pushed away from the fissure, swimming hard toward the dim, fractured light of the cave entrance. Every muscle in Avery's body was coiled, her eyes tracking the walls as they trembled. She realized then that their mission had irrevocably changed. They weren't just curious teenagers documenting a decline. They were witnesses to a slow-motion environmental crime scene, and the ticking clock was now deafening.

As they burst through the surface into the main basin of the spring, the bright Florida sun hit them like a physical blow. The water here was barely waist-deep, a pathetic fraction of its usual glory.

Avery hauled herself onto the wooden boardwalk, gasping for breath, her hands shaking as she clicked off her headlamp. She looked back at the water. It was quiet. Too quiet.

"We have everything we need," Declan said, surfacing behind her and pulling his camera unit close to his chest like a holy relic.

"Is it enough?" Avery asked, staring at the darkened water.

Micah's voice returned to the comms, breathless and urgent. "I've matched the nitrate signatures, guys. It's an exact match to the fertilizer logs from the Enclave. And the thermal data? It's consistent with the cooling capacity of a development that size. You have the physical, chemical, and historical evidence. Now, we have to make sure they can't ignore it."

Avery looked toward the distant trees, where the rooftops of the Enclave development peeked over the canopy. "They won't be able to ignore it," she said, her voice hardening. "Because we're going to make sure the whole town sees what's hiding in the dark."

She stood up, ignoring the ache in her limbs, and turned toward the parking lot where Zara was waiting. The investigation was over; the confrontation was beginning.

Chapter Twelve

Nitrate Traces

The air inside the research shed was stale, tasting of ozone and old dust, but the hum of the portable water-quality meter was a frantic heartbeat in the silence. Micah gripped the device like a lifeline, his knuckles white.

"The baseline for this sector should be point-zero-four milligrams per liter," Micah whispered, his voice trembling with a forced, fragile calm. "It's a fragile balance, Avery. The eelgrass needs the low-nutrient flow to stay clear. If the nitrates go up, the algae chokes out everything."

Avery leaned over his shoulder, her eyes fixed on the digital display. The numbers crawled upward, twitching like a dying insect.

0.12.

"That's three times the normal limit," she said, her voice dropping into that quiet, analytical register she used when observing the spring's currents. "And we're fifty yards upstream from the main tourist basin. This isn't runoff from the public parking lot."

Micah looked at her, his expression shattering. The gentle, mediator-personality he usually wore—the one that smoothed over conflicts at the park gate—was being replaced by a cold, hard indignation. "It's not just a fluctuation, Avery. It's a signature. This is synthetic fertilizer.

High-concentration lawn and decorative turf grade. Nobody living in the park residential zone uses this much at once."

Avery grabbed her field notebook, clicking her pen with a sharp, rhythmic snap. "Move to the secondary channel. The one that cuts behind the old citrus grove barrier."

They pushed off from the bank, their kayaks gliding over the unnaturally shallow water. The spring, once a vibrant, rhythmic pulse of life, felt lethargic, the water sluggish and murky. Avery watched the bottom, her heart sinking. Where there should have been healthy, swaying ribbons of *Vallisneria*, she saw patches of dark, fuzzy sludge. The vegetation was gasping.

She dipped her paddle, feeling the water's resistance. It didn't have the crisp, cold bite it usually possessed. It felt heavy, tempered by warmth and something foreign.

"Stop," Avery hissed.

She steered her kayak into a tangle of cypress roots, pulling a vial from her kit. She reached into the water—not into the main flow, but into a stagnant pocket near the bank where the water should have been clear, filtered through the limestone labyrinth. Instead, she saw a film of iridescent oil clinging to the edges of the sediment.

Micah hovered beside her, his kayak bumping hers. "Avery, look at the bank."

She followed his gaze. The roots of a massive, ancient bald cypress were exposed, the soil beneath them stripped away by recent, aggressive erosion. But it wasn't just soil. There were jagged, pale shards of concrete and plastic piping protruding from the mud, remnants of a drainage system that clearly hadn't been mapped in any official park document.

"That's not from the park maintenance department," Micah said, his voice dropping to a low, dangerous register. "That's industrial-grade PVC. It's been buried here for months, maybe years, hidden under the silt."

Avery took the sample, the water in the vial cloudy and amber-tinted. She held it up to the dappled sunlight filtering through the canopy. "It's not just a drop in the basin, Micah. It's an infection. Someone is using these old, defunct drainage arteries as a waste pipe."

"If we follow the line of these drainage markers," Micah said, his eyes scanning the swampy terrain with a newfound, sharp focus, "we can trace where it originates. But look at the water quality reading again."

He held the meter out.

0.85.

Micah's breath hitched. "This is catastrophic. This isn't just 'runoff.' This is a direct injection of chemical waste into the aquifer's recharge zone. I've spent my life telling people that the spring is a living, breathing thing. I thought I was being poetic. I didn't realize I was being literal."

Avery felt a cold shiver trace the line of her spine. She wasn't just a photographer anymore; she wasn't just a girl who knew the currents. She was a witness to a crime against the very land that defined her. She looked at the murky water, then at the dying eelgrass, and finally at Micah. The naturalist was gone, replaced by a guardian who looked ready to dismantle the world to fix what was broken.

"The Enclave," Avery said, her voice steady. "They said they were 'landscaping' the buffer zone last year. They paved over the natural wetlands to make room for their 'luxury aesthetic.' They didn't just remove the trees, Micah. They moved the water."

Micah looked down at the probe, the number blinking insistently: *0.92.*

"They didn't just move the water," he replied, his eyes hardening. "They turned it into a disposal system. And if this hits the main cavern system, it won't just kill the grass. It will compromise the entire structural integrity of the basin. The limestone will start to dissolve faster under these chemical conditions. We're not looking at a dry spring anymore. We're looking at a collapse."

Avery dipped her paddle again, deeper this time, finding the strength to push against the sluggish, poisoned current. "Then we don't go back to the shed. We follow the pipe as far as the current will take us, until we find the source of that nitrogen spike."

Micah looked toward the Enclave's manicured, artificial lawns—a bright, neon-green stain on the horizon that stood in stark contrast to the wild, suffering beauty of the spring. He nodded once, a gesture of grim commitment.

"Let's see where the poison leads."

Micah knelt at the edge of the secondary channel, his movements practiced and rhythmic. He didn't just dip the test kit; he cradled it, his hands submerged in the cool, shallow water. Beside him, Avery sat cross-legged on a mossy limestone shelf, her camera lens capped, her gaze fixed on the struggling patches of *Vallisneria*—eelgrass—that swayed listlessly in the sluggish current.

"The numbers aren't making sense," Micah said, his voice barely a murmur above the gentle lap of water against the rock. He held the test strip up against the color-coded vial. The nitrate indicator was a sickly, deep magenta. "We're in the middle of a protected basin, Avery. This isn't just a seasonal spike. This is industrial-grade contamination."

Avery leaned in, squinting at the strip. "Magenta. That's... that's off the charts for this time of year, isn't it?"

"It's not just off the charts," Micah replied, his usual calm demeanor fraying at the edges. He sat back on his heels, wiping his damp hands on his jeans. "It's a death sentence for the vegetation. Look at the eelgrass."

Avery shifted, pointing a slender finger toward the riverbed. "I've been watching it all morning. It's not just the color, Micah. Look at the base of the stalks. They're necrotic. They aren't getting smothered by silt; they're being chemically burned. The spring isn't just running low—it's being poisoned."

She felt a cold, hard knot form in her stomach. For weeks, they had treated this like a mystery to be solved, a puzzle of flow rates and subterranean geography. But seeing the decay firsthand—the way the vibrant green of the spring was being bleached into a dull, brownish grey—shifted the stakes. This wasn't a crime of apathy; it was an act of violence against the landscape she had called home her entire life.

Micah stood up, his face set in a grim, uncharacteristic line. He moved toward the drainage marker—a rusted metal grate obscured by a tangle of invasive vines near the park's perimeter. "The runoff has to be coming from somewhere specific. The way the concentration pools here, near the northern intake... it means it's not leaching in from the whole park. It's being directed."

He pulled his phone, accessing the rough, hand-drawn map Declan had compiled. "If the Enclave's irrigation system is pulling from the primary aquifer, it's creating a vacuum. That vacuum is pulling surface water—contaminated surface water—down through the limestone fissures instead of letting it filter naturally."

Avery stood and joined him at the grate. She reached down, pulling back a handful of brittle, dead ferns. Beneath the rot, the soil was discolored—a strange, chemical orange that crusted over the limestone. "Look at the patterns in the sediment," she whispered. "This isn't organic. This is fertilizer runoff. And look there."

She pointed to a series of PVC pipe remnants half-buried in the mud, leading away from the park's boundary toward the manicured, thirsty lawns of the new development bordering the property.

"They aren't just pumping out," Micah said, the realization landing with a heavy finality. "They're flushing back. They're using our sinkholes as storm drains."

The silence that stretched between them was heavy, broken only by the distant, mocking chirp of a cardinal. Micah's hands were shaking, not with fear, but with a quiet, simmering indignation. He had spent his life acting as the bridge between people and the park, the mediator who believed that if people just understood the ecosystem, they would protect it. But this? This was premeditated.

"We need to document the connection," Avery said, her voice dropping into that focused, clinical tone she used when she was tracking wildlife. She pulled her camera from her bag, but instead of checking the settings for light or composition, she began snapping photos of the discharge pipes, the discolored soil, and the dead vegetation. "We need to show exactly how the 'Enclave' is using the park's own geology to hide their waste. If we can prove the path of the contamination, we take away the 'seasonal fluctuation' excuse."

"It goes deeper than that," Micah noted, his eyes scanning the ridgeline where the luxury development's roofline peeked over the cypress canopy. "If they're pumping that much volume, the sinkhole structure is already under extreme pressure. We're not just looking at pollution, Avery. We're looking at a potential collapse of the aquifer conduit."

He turned back to her, his gaze intense. "If we go to the Town Hall with just 'it's dirty,' they'll laugh us out of the room. We need to tie the land-use history—the old grove maps, the permit logs Zara pulled—directly to the chemical toxicity we're holding in our hands right now."

Avery felt a surge of resolve. The quiet girl who watched from the banks of the spring was gone, replaced by someone who understood that silence was a complicit act. "I have the visual documentation of the vegetation decay over the last three weeks," she said, tapping her camera. "You have the chemical evidence. Zara has the paper trail. We don't just have a story anymore, Micah. We have a case."

Micah reached out, placing a hand on the rusted drainage marker. He didn't try to clear the vines anymore. Instead, he pulled out his notebook and began mapping the runoff's trajectory, tracing the invisible lines of water as they snaked beneath the earth, cutting through the foundations of their community.

"They thought the spring was a passive resource," Micah said softly, his eyes reflecting the dark, sluggish water. "They thought it was something they could just drain and fill. They forgot that everything is connected. If we show them the trail, they can't pretend they didn't see where the water leads."

Avery looked toward the main basin, where the water usually shimmered with a turquoise intensity. It looked tired today, strained under the weight of the invisible sludge seeping into its veins. "Let's get back to the shed," she said, her voice steady. "We need to cross-reference these nitrate spikes with the irrigation schedules Zara found. If we show a direct correlation between their watering times and the flow decline, we can prove intent."

As they turned to leave, the breeze shifted, carrying the faint, chemical tang of high-nitrogen fertilizer. It was a smell that didn't belong in the cypress woods—a sharp, metallic intruder in the humid sanctuary of the spring. Avery took one last look at the dying eelgrass, the picture of the crisis captured in her mind as vividly as it was on her memory card.

The path back to the research shed felt longer, the humid air pressing in on them, but the uncertainty that had haunted their early days of investigation was gone. They knew the 'how' and the 'why.' Now, it was just a matter of proving it to the people who had been convinced that Glasswater Spring was a bottomless well, rather than a living, breathing system that was gasping its last breaths.

"We're going to need to present this clearly," Micah said, breaking the silence as they reached the park trail. "No jargon. No vague theories. Just the raw, ugly truth of what happens when you decide that a community's heart is just another line item in a development plan."

Avery nodded, her stride lengthening. "Zara will know how to frame it. She's been waiting for this exact kind of evidence."

"We're not just saving the water," Micah added, looking back at the trees one last time. "We're saving the future of the whole region. If the conduit collapses, the sinkhole isn't just a park problem. It's a town-wide disaster."

The urgency of the moment finally hit her—not as a fear, but as a driving force. Every step toward the shed felt like a step toward a reckoning. The investigation had outgrown the park, outgrown their small group, and was now hurtling toward a confrontation that would decide if the Glasswater Spring remained a natural wonder or became a forgotten relic of a community's hubris.

Chapter Thirteen

THE DRONE'S EYE VIEW

The northern perimeter of Glasswater State Park was a tangle of saw palmetto and scrub oak, a place where the manicured boardwalks gave way to the brutal, chaotic reality of the Florida wild.

Declan adjusted his harness, the heavy pack containing the drone's docking station pulling at his shoulders. He didn't stop moving until they reached the limestone ridge—a jagged spine of grey rock that overlooked the forbidden zone behind the new Enclave development.

Avery kept pace, her eyes constantly scanning the ground. She noticed the soil here was wrong. Instead of the damp, dark humus that should have been underfoot, the earth was parched, gray, and cracked, like the floor of a kiln. She reached down, brushing away a layer of fine, powdery dust. "Declan, look at this. The ferns—they're brittle. They haven't had a proper soak in weeks."

Declan nodded grimly, already setting the drone down on a flat, sun-bleached rock. "It's not just the surface, Avery. The water table isn't just low—it's being actively pushed away."

He tapped his tablet, the drone's rotors giving a high-pitched, angry whine as they spun to life. He lifted it above the canopy, the machine disappearing into the shimmering heat haze that hovered over the Florida basin.

Avery leaned in, squinting at the primary screen. The drone fed back high-definition infrared footage. At first, it looked like a standard aerial survey of forest canopy, but as Declan tilted the camera, the perspective shifted.

"There," Avery whispered, her finger tracing a dark, artificial scar on the ground below.

The screen displayed a massive, concrete-lined drainage basin hidden behind a curtain of pine trees, just yards beyond the park boundary. It was designed to channel water away from the Enclave's manicured lawns and golf course. But it wasn't just a drain.

"They didn't just route the runoff into the wetlands," Declan said, his voice hard. "Look at the junction. See that piping? They've drilled directly into the porous limestone shelf. They're dumping untreated runoff, full of sediment and whatever chemical cocktails they use on those greens, directly into the subterranean conduits that feed the Grotto."

Avery felt a chill that had nothing to do with the humidity. She zoomed in on the feed. The footage clearly showed a pressurized, industrial-grade pump system forcing water through a series of secondary limestone fractures.

"They're siphoning," Avery realized, her voice trembling. "They aren't just draining the stormwater; they're using the karst system as a vacuum. They're drawing the natural flow of the spring toward their own cisterns to keep their landscaping lush while the spring itself dies of thirst."

Declan flew the drone lower, hovering near a cluster of sinkholes. The feed showed the ground sagging, the soil slumped into deep, unstable funnels.

"The architecture of the spring is shifting," Declan said, his hands dancing across the controls. "By forcing the water away, they've created a vacuum in the limestone layer. It's causing the internal structures to collapse. That explains why the flow has been so erratic. The water isn't just low—an artificial pressure gradient is rerouting it. It's an artificial heart attack."

Avery watched the screen, the reality of the image hitting her like a physical blow. The drone panned over the entire northern section, revealing a complex, spider-web network of pipes and diversions that completely bypassed the natural filtration zones of the park. It was a masterpiece of engineering, designed with a complete, cold disregard for the geology beneath it.

"They didn't make a mistake," Avery said, her gaze fixed on the monitor. "They calculated this. They knew exactly how much pressure the karst could take before it broke."

"If we present this, it's not just an argument about environmental aesthetics anymore," Declan replied, hovering the drone to capture a final, damning wide shot of the entire diversion scheme. "This is a structural liability. It's evidence of negligence that could threaten the foundations of their own houses, let alone the park."

He began the landing sequence, the drone descending through the canopy with a rhythmic thrum. Avery didn't look away from the screen, even as the image flickered with interference from the dense foliage.

"Declan," she said, finally turning to him. Her eyes were sharp, the quiet observer replaced by someone who now saw exactly how the world worked. "We have the footage. We have the proof that this isn't natural. But how do we explain this to the town? They'll see these blueprints and see 'progress.' We have to make them see that if they kill the spring, they kill the foundation of everything they've built."

Declan snapped the drone into its case, the silence of the woods returning, heavy and expectant. "We show them the autopsy, Avery. We show them the maps, the chemical spikes, and now this—the smoking gun. We connect the history Zara is digging up with the physics we're looking at right now. It's not just a story anymore. It's a closed-loop system of consequences."

Avery stood up, her legs feeling heavy, grounded by the weight of the realization. She looked toward the spring, hidden somewhere behind the dense, green wall of trees. The air felt thin, strained.

"We need to get back to the others," she said. "We have the final piece. Now, we have to make sure they're ready for the fallout."

They moved back through the undergrowth, moving faster now, the urgency of their discovery driving them toward the trailhead. The sun was dipping lower, casting long, fractured shadows across the park, but for the first time, the heat didn't feel like a burden. It felt like a countdown.

The hum of the drone's rotors was a thin, high-pitched whine that seemed to slice through the oppressive humidity of the northern perimeter. Declan stood on a slight rise of limestone, his brow furrowed as he manipulated the handheld controller. His eyes flickered between the small monitor and the dense, scrub-oak canopy ahead, his movements precise, almost surgical.

Avery crouched beside him, her phone plugged into the receiver, the display showing a jagged, bird's-eye view of the landscape. She wasn't looking at the trees; she was looking at the ground. She was looking at the way the land *didn't* look like it should.

"Bring it left," Avery said, her voice low. "The drainage easement—the one on the old county map—it's supposed to terminate near the limestone shelf. Look at the silt patterns."

Declan nudged the joystick. The drone banked sharply, the camera stabilizing to show a swathe of land that had been cleared for the Enclave expansion. It was a scar of raw earth, a jagged wound of red clay and construction debris that bled into the park's northern boundary.

"See that?" Declan pointed to a series of dark, rectangular concrete basins tucked behind a row of privacy fencing. "Those aren't just retention ponds.

Look at the elevation profile. The water isn't draining south toward the aquifer recharge zone. They've graded the entire site to divert it toward the northern sinkhole cluster."

Avery felt a chill that had nothing to do with the morning mist. She tapped the screen, zooming in. "If they're pumping the runoff into those sinkholes, they aren't just flooding the caves. They're dumping untreated site water—oil, fertilizer, industrial-grade silt—directly into the throat of the spring's main conduit."

"It's not just dumping, Avery," Declan said, his tone turning sharp, the thrill of the explorer replaced by the grim clarity of the scientist. "Look at the flow velocity in the thermal overlay."

He toggled the sensor view. The monitor shifted into a spectrum of blues and reds. A concentrated, burning orange stream was bleeding from the construction site, plunging into the dark, cold blue of the limestone cavity.

"They aren't just blocking the spring's natural recharge," Avery whispered, the realization hitting her with the force of a physical blow. "They're suffocating it. They've clogged the lungs of the entire system."

She moved her finger across the screen, tracing the invisible path of the water deep beneath their feet. The geology of the park wasn't static; it was a living, breathing network of veins. What happened here, at the periphery, was being echoed at the spring head, miles away, where the water had slowed to a lethargic, muddy crawl.

"It's not a drought," Declan muttered, his thumbs hovering over the controls. "It's a slow-motion collapse. They've rerouted the flow to prevent the Enclave from flooding, and in doing so, they've cut off the pressure that feeds the Glasswater Basin."

Avery pulled her phone away, pacing the small, uneven patch of ground. She thought of the spring yesterday—the way the manatees had lingered in the shallows, confused by the lack of clear, cold water. She thought of the way the native vegetation was already showing signs of stress, the vibrant greens turning a brittle, bruised yellow.

"We have to link this," she said, her voice gaining a frantic edge. "If we just take this footage to the Town Hall, they'll call it an accident. A byproduct of 'necessary development.' We need to show them the math. We need to show them the history."

"Zara's digging into the permits," Declan reminded her, his eyes still glued to the drone's feed. "If she finds that the Board of Trustees knowingly signed off on a grading plan that contradicts the park's original conservation mandate, we have more than just a scientific problem. We have a legal one."

Avery looked back at the screen. The drone was hovering over a specific vent, a deep, jagged fissure in the rock. It looked like a black eye in the middle of the woods. "This isn't just about the spring anymore, Declan. This is about the decision to prioritize a housing development over the very water table that makes this whole region livable. They've traded our foundation for a few luxury square feet."

Declan brought the drone into a hover, the rotors whining as he prepared for the final sweep. "Do you think it's enough? Can we actually walk into that meeting and pull the rug out from under them?"

Avery looked toward the distant, hazy tree line where the spring resided. She could almost feel the heartbeat of the earth—a rhythm that had been steady for centuries, now skipping beats, struggling to maintain its cadence. She thought of the community members who looked at the spring as a park, a place to walk the dog or take a dip on a Sunday, unaware that the ground beneath them was failing.

"It has to be enough," Avery said, her voice steadying. "Because if it's not, this isn't just the end of the spring. It's the end of everything that draws water from this aquifer. The contamination is going to hit the wells next. It's not just an environmental issue, Declan. It's a crisis of public safety."

She knelt back down, taking the receiver from him. "Take it up to five hundred feet. I want to see the whole catchment area. I want to see the extent of the grading."

As the drone climbed, the view expanded. The screen captured a terrifyingly wide perspective—a web of roads, driveways, and concrete foundations that acted as a giant, impermeable lid, preventing rainwater from ever reaching the soil. The scope of it was overwhelming. It wasn't one bad actor; it was a systematic erasure of the landscape.

"Look at the drainage patterns," Avery pointed out, tracing the way the natural landscape had been leveled into artificial slopes. "They've turned the entire hillside into a concrete slide. Any rain that hits this area is being shunted straight into that sinkhole. They've essentially weaponized the surface runoff to keep their development dry."

Declan let out a long, low whistle. "It's an engineering marvel, I'll give them that. And a complete ecological death sentence."

"We need to document the GPS coordinates of every point of diversion," Avery said, her mind already organizing the data into a narrative. "When we present this, it can't be emotional. It has to be a map of their negligence. If we lay out the timeline—the permits Zara finds, the chemical spikes Micah is documenting, and this—the physical redirection of the water—it's going to be impossible to ignore."

She felt a strange sense of calm settling over her. The quiet observer had finally seen enough. The mystery was gone, replaced by the ugly, undeniable truth of the mechanism. She wasn't just a girl who knew the spring; she was a witness to its murder.

"Record it," she commanded softly.

Declan initiated the high-resolution capture. The drone tilted its nose, sweeping over the Enclave, the camera capturing every pipe, every culvert, every deliberate choice made by architects who had never bothered to look at what lay beneath the glass.

"This is going to be a long night," Declan said, watching the progress bar on his controller.

"Good," Avery replied, looking up at the sky, where the sun was beginning to dip, casting long, sharp shadows across the ruined landscape. "The truth is going to be much longer."

She watched the drone circle, a tiny, buzzing sentinel above a landscape that had been betrayed. The weight of the evidence felt heavy in her pocket, a small hard drive full of maps and telemetry that could change the fate of their home. There was no more guessing. There was only the work.

"Let's get this back to the shed," Avery said. "Micah and Zara are going to need to see this, and we have to synthesize it all before the doors open tomorrow. We're not going to ask for their help, Declan. We're going to show them why they have no choice but to provide it."

As Declan maneuvered the drone back toward their position, Avery turned her back on the construction site and started walking toward the tree line. She walked with a new, deliberate pace. Her connection to the spring was no longer a quiet, private devotion; it was an active defense. The water was dying, but she had the proof of its heartbeat, and she was going to fight to bring it back to life.

"Wait," Declan said, pausing as the drone landed in the tall grass. He looked at her, his eyes searching hers. "You know, once we walk in there with this, there's no going back. This puts a target on us, too."

Avery didn't break her stride. "The spring has been holding us up for a long time, Declan. I think it's our turn to hold it up."

Chapter Fourteen

RECORDS OF THE OLD GROVE

The air in the County Clerk's archive basement tasted like slow-decaying paper and pulverized limestone. It was a space that felt entirely disconnected from the humid, sun-drenched vibrancy of the spring, a tomb of bureaucracy where the only thing flowing was the hum of a failing fluorescent light fixture overhead.

Zara adjusted her glasses, the bridge of her nose aching from three hours of squinting at microfiche and crumbling parchment. Her fingers were stained grey with archival dust. She wasn't looking for a story anymore; she was looking for a weapon.

She unrolled a map from 1954, its edges brittle as dried tobacco leaves. Beside it, she laid out the glossy, color-coded zoning printouts she'd pulled from the planning department's digital portal that morning. The transition was sickening. Where the old map showed a sprawling, permeable swathe of palmetto scrub and recharge wetlands—the very heart of the aquifer's natural "breathing" space—the modern overlay showed the Enclave's dense grid of concrete pads, high-density pavement, and tiered luxury homes.

"Got you," she whispered, her voice sounding raspy in the stillness.

She traced a line with a graphite pencil. The 1954 map had a handwritten annotation in red ink in the margin: *Restricted. Primary recharge zone. High permeability.*

Then, she pulled up the 1998 land-use change request. It was a dense stack of legalese buried behind a clerical error that had mislabeled the area as "disturbed industrial wasteland." She flipped to the signature page. The developer hadn't just built on top of the aquifer; they had systematically dismantled the natural filtration pathways, filling in the ephemeral ponds with construction debris and capping the limestone sinkholes to provide "stable foundations" for the clubhouse.

It wasn't a drought. It was an extraction. They had paved over the lungs of the spring, and the current, smaller development was just the final, suffocating stitch in a wound that had been opening for thirty years.

Zara grabbed her phone, her thumb hovering over the group chat. She hesitated, looking back at the juxtaposition of the maps. The administrative silence of these documents was louder than any environmental protest she'd ever covered. These weren't just zoning permits; they were blueprints for a slow-motion disaster, signed off by officials who had long since retired, leaving behind a legacy that now fell squarely on their shoulders.

She took a rapid-fire series of photos of the documents, the flash illuminating the dust motes dancing in the stagnant air. She didn't need to write an article anymore. She had a roadmap of culpability.

The sun was dipping low over the northern ridge, casting long, bruised-purple shadows across the abandoned citrus grove. The trees here were skeletal, their bark peeling like sunburned skin, leaning away from the encroaching weeds.

Micah stood near the edge of the irrigation ditch, his boots caked in the reddish clay that distinguished this part of the basin. He was knee-deep in a cluster of invasive ferns that had thrived where the native flora had withered. When he heard the crunch of gravel behind him, he didn't turn. He knew the precise, hurried cadence of Zara's walk.

"I found it," Zara said, her voice tight with a mix of triumph and cold anger. She didn't offer a greeting; she went straight to the satchel, pulling out the printouts and the grainy photos she'd snapped at the archives.

Micah stepped out of the brush, taking the papers from her. He looked at the 1954 map, then squinted toward the horizon where the glint of the Enclave's roofline broke the treeline.

"They didn't just build houses," Micah said, his voice barely a whisper as he scanned the technical annotations on the map. "They re-engineered the entire water table. Look at this—they diverted the runoff away from the natural sinkholes and funneled it directly into the retention ponds for the golf course."

"The water isn't going into the aquifer anymore," Zara confirmed, pointing to the modern overlay. "It's being treated as waste. It's being piped away, managed, and lost. They've turned a natural sponge into a waterproof roof. No wonder the spring is dropping. The rain is not feeding it; the development is starving it."

Micah knelt, pushing aside the leaf litter near the irrigation ditch. He pulled a small probe from his pack and inserted it deep into the soil. He watched the digital readout climb. "High concentrations of phosphate and synthetic nitrates," he muttered. "The same fertilizers we saw in the Grotto. They're washing the lawns, and because they paved the recharge zone, the water doesn't filter through the sand. It hits the limestone shelf and slips straight into the conduit—unfiltered, acidic, and toxic."

He looked up at Zara, his face pale. "It's a double hit. We're losing the volume of the flow because of the paving, and we're poisoning what's left of the water that actually reaches the spring."

"It's not just an environmental issue, Micah," Zara said, her journalistic instinct shifting into a more aggressive gear. "This is a legal one. They obtained these permits under the guise of 'remediation.' They claimed they were cleaning up a brownfield site, but they used that as a loophole to bypass the environmental impact assessments required for building on a primary recharge zone. They lied to the county, and the county didn't bother to check because the developer's 'remediation' plan sounded cheaper than the alternatives."

Micah stood up, his eyes sweeping over the grove. The silence here wasn't peaceful; it felt hollow, like a lung that couldn't quite inflate. "We've been looking at the spring as if it were a sick patient," he said. "We've been trying to treat the symptoms, but the history of this land is the disease."

"The Town Hall meeting is in two days," Zara said, tucking the documents securely into her bag. "We have the maps, we have the chemical evidence, and we have the paper trail that shows exactly who signed off on the death warrant of this spring. It's not just teenagers yelling about dry grass anymore. It's an exposure."

Micah looked toward the spring, hidden somewhere beneath the canopy of oaks and cypress. "Avery and Declan are expecting us at the shed. If we're going to do this, we have to make the community see that this isn't just about the water. It's about the fact that they've been sold a lie about the safety and sustainability of their own homes."

Zara nodded, her eyes hardening. "They want to talk about progress? Fine. Let's talk about what that progress cost. Let's show them exactly what's drying up under their feet."

As they began the hike back toward the research shed, the sky turned a deep, blood-orange, silhouetting the dead trees of the grove. The path felt different now—less like a trail through nature and more like a line drawn through a crime scene. Every step they took, every patch of paved road that encroached on the wild, felt like a piece of evidence they were cataloging.

They weren't just observers anymore; they were the ones tasked with bringing the truth out from the dark of the archives and into the unforgiving light of public scrutiny. The spring was waiting, and for the first time, they knew exactly how to fight for its return.

The rusted barbed-wire fence groaned as Zara squeezed through a gap, her notebook clutched to her chest like a shield. Beyond the boundary, the landscape shifted abruptly. Where the park ended in lush, damp ferns, the abandoned citrus grove began—a graveyard of gnarled, skeletal trees that looked as if they'd been clawed into shape by the very soil that refused to feed them.

Micah was already there, kneeling in the dirt near a dried-out irrigation trench. He didn't look up as Zara approached, his fingers sifting through a handful of soil that crumbled into a fine, grey dust instead of clumping like healthy Florida earth.

"It's not just the drought, Zara," Micah said, his voice stripped of its usual warmth. "It's the soul of the place. It's been bleached."

Zara crouched beside him, spreading out the oversized, brittle map she'd liberated from the Clerk's archives. She flattened it against the hard, sun-baked ground, weighing down the corners with chunks of limestone. "I found it. The smoking gun isn't a pipe or a pump. It's a permit signed in 1974, buried under three different property transfers. They rezoned this specific sector as 'Industrial-Agricultural Overlay' despite the hydrological reports explicitly flagging the recharge potential. Look at the lines."

She traced a finger over the faded ink. "They didn't just plant trees here. They paved over the natural sinkhole catchment that feeds the upper conduit. It's like they put a plastic bag over the spring's lungs fifty years ago and just waited for it to suffocate."

Micah leaned in, his eyes scanning the intersection of the hand-drawn map and the reality before them. He pointed to a low-lying depression near a rotted orange stump. "That's where the runoff from the old chemical sprays collected. If you look at the soil profile, the compaction is absolute. Water can't penetrate this crust; it just sheets off into the storm drains of the Enclave now, instead of filtering down into the Floridan Aquifer."

He stood up, walking toward a patch of stunted saw palmetto that looked brittle enough to shatter. "We've been looking for a villain in the present—a greedy developer, a malfunctioning valve. But this? This is intergenerational malpractice. The citrus growers paved the top, the developers built the houses, and the city board kept the permits flowing because no one wanted to admit the 'development' was actually a slow-motion environmental suicide."

Zara felt a cold prickle of realization at the base of her neck. She tapped the map, her pen hovering over the red-lined perimeter of the Enclave. "That's why the Board is fighting us so hard. It isn't just about the new project. If they admit the spring is failing because of the recharge zone being compromised, they have to re-evaluate the entire history of land use in this county. They'd have to admit that the property tax revenue they've been relying on for decades is built on a foundation of ecological destruction."

"They don't want the truth," Micah whispered, looking out toward the shimmering heat haze where the park's canopy met the stark, artificial lines of the new construction. "They want the profit to outlast the memory of what this place used to be."

Zara stood, dusting the soil from her knees. She felt the weight of the documents in her bag—not just paper, but the forensic timeline of a crime. "They think we're just kids looking at a pond. They think if they wait for the next rainfall, we'll get bored, and the water will come back on its own, hiding their mistakes under an inch of rain."

Micah picked up a piece of the desiccated limestone near his feet. It was porous, meant to hold water, but it felt as light and hollow as a dried bone.

"The rain won't save us. Not if the path to the aquifer is sealed. We have to show them exactly how the geometry of this place was broken."

"I have the permits," Zara said, her voice hardening with a new, professional edge. "And Declan has the sonar maps that show the specific pressure of those industrial-grade pumps is diverting the flow. If we put your nitrate samples from the runoff right next to the archival records of the chemical use in this grove, it creates a narrative they can't dismiss as teenage alarmism."

"It makes it a legacy problem," Micah added, a grim smile touching his lips. "It makes it a liability for the county."

Zara looked at the vast, ruined landscape—the silent trees, the cracked earth, the ghost of a water system that had kept the community alive for centuries. She felt a profound shift in her own perspective. She had come here seeking a scoop, a headline, a story. But as she looked at the map, she realized this wasn't a story to be told; it was a wrong that had to be righted.

"We need to get this to the Town Hall," Zara said, rolling the map tightly. "We don't just present the facts. We present the bill. We show them exactly what their 'progress' has cost us."

Micah looked back at the park. "Avery will know what this means for the grotto. When she sees how the surface drainage is linked to the cave systems Declan mapped, she'll be able to explain it better than any of us. We have everything we need to dismantle their excuses."

As they began the trek back toward the spring, the atmosphere felt different. The humidity still clung to them, and the sun still beat down with indifferent intensity, but the crushing sense of helplessness had vanished. In its place was the cold, sharp clarity of a well-organized case.

"Zara?" Micah asked as they reached the edge of the woods.

"Yeah?"

"Do you think they'll even listen? Or are we just walking into a brick wall?"

Zara stopped, turning to look back at the skeletal remains of the grove. She could see the faint, shimmering line of the Enclave's manicured lawns in the distance—a bright green bruise on the Florida landscape. "They don't have to listen to us because they like us, Micah. They have to listen because I'm going to lay it out so clearly that the taxpayers are going to start asking why their property values are tied to a dying aquifer. Fear is a better motivator than empathy, and I have the receipts to make them very, very afraid of the liability."

She didn't sound like a high school journalist anymore. There was a flinty, uncompromising edge to her tone that Micah had only seen once or twice before.

"Let's get back," he agreed, his own apprehension replaced by a steady, grounding resolve.

They emerged from the treeline just as the sky began to transition into the bruised violet of a Florida dusk. The park was quiet, the usual cacophony of weekend tourists and kayakers long gone, leaving the spring to reflect the first faint stars.

Avery and Declan were already huddled over the folding table at the research shed. The glow of a laptop screen illuminated their faces—Avery, looking exhausted but intensely focused, and Declan, who was meticulously overlaying sonar data onto a topographical grid.

"You're late," Declan said without looking up, his voice tight. "But tell me you have something good. The flow rate in the North Grotto just dipped another three millimeters. If we don't hit the hydrostatic threshold by tomorrow morning, the upper shelf is going to be permanently exposed to air. We're talking about irreversible calcification of the biofilm."

Avery looked up, her eyes wide. She'd been staring at the water for so long that her pupils seemed to hold the dark, bruised color of the pool. "The fish are starting to cluster at the boil, Declan. They're looking for the deepest water, but the boil is moving. It's shifting away from the main vent."

Zara stepped forward, tossing the rolls of vellum and the stack of archive photos onto the table. They landed with a heavy, final sound. "I have more than something good. I have the architect's signature on the death of the spring."

Declan stopped clicking. He pushed his glasses up his nose, his gaze flickering from the maps to Zara, and then to the document she was unfurling.

"This isn't just an old map," he murmured, his geologist's brain instantly translating the lines. "This is a drainage projection from '74. Look at the grading—it was designed to funnel everything into the western basin, away from the natural sinkholes."

Avery stood up, leaning over the table. Her fingers traced the red ink on the old document. "I remember my grandfather talking about this area," she whispered. "He said the spring used to 'breathe' during the rainy season. That the water would rise and fall through the limestone vents. I thought he was just being poetic."

"He wasn't," Micah said, stepping up behind her. "He was describing the natural cycle of an uncompromised aquifer. The construction didn't just stop the flow; it choked the breath out of it."

Declan grabbed a laser pointer and flicked it on. He projected a bright red dot onto the map, tracking the flow of the water from the Enclave's artificial retention ponds, across the paved roads, and right past the park's entrance. "The path of least resistance. They didn't build for longevity; they built for convenience. They rerouted the surface recharge into these storm pipes, which means that instead of filtering through forty feet of sand and porous limestone, the rainwater—loaded with pesticides from the lawn maintenance and motor oil from the streets—is being dumped into the municipal sewer system."

Avery stared at the map, the logic slotting into place like a gear in a machine. "And when the water isn't there to recharge the aquifer," she said, her voice growing steady, "the pressure in the main conduit drops. That's why the Grotto is drying up. It's not just that we aren't getting rain; it's that the water we *are* getting is being diverted away from the source."

"It's a design flaw," Declan said, a savage sort of excitement in his voice. "A massive, structural, multimillion-dollar mistake. And they've been hiding it behind maintenance logs and 'drought management' pamphlets for decades."

"Tomorrow at the meeting," Zara said, her voice dropping to a low, dangerous register, "I'm not just going to talk about the fish dying. I'm going to talk about the structural integrity of the entire neighborhood. If the foundation of the Enclave is built on a porous, 'disturbed' zone that was supposed to be a primary recharge area, they're sitting on a potential sinkhole. The county isn't just looking at a dried-up spring—they're looking at a multi-million dollar liability."

Avery looked at the group—the journalist, the geologist, the naturalist. They were four teenagers in a shed, surrounded by the smell of damp earth and the distant, lonely sound of the spring's failing pulse. But as she looked at the map, she didn't feel like a victim of the situation anymore. She felt like a curator of a secret.

"We have to be careful," Avery warned, though there was a new steel in her spine. "The Board isn't going to roll over just because we show them a map."

"They won't have to roll over," Zara replied, her thumb hovering over the 'publish' button on her tablet, where she had already drafted the skeleton of her report. "They just have to realize that keeping the secret is more expensive than telling the truth."

The room fell silent. Outside, the crickets began their rhythmic chirping, a sound that usually lulled the park to sleep but now sounded like a countdown.

"We need to test the sediment in the detention ponds," Declan said, breaking the silence. "If we can prove that the toxic load from the Enclave is the same as the nitrates we're seeing in the Grotto, we have a chemical link to the physical rerouting. It's undeniable."

"I'll go," Micah said immediately. "I have the collection kits. I can get samples before dawn."

"I'll come with you," Avery said. She looked at the spring, her face set. "I need to see the vent levels once more at high tide. If the flow rate is what I think it is, the water is going to pull back another ten feet tonight."

"And I," Zara said, reaching for her notebook, "am going to spend the next eight hours writing the most uncomfortable article this county has ever seen. I want every board member's name on the permits for the last thirty years. I want to know who signed what and who got paid for it."

Declan looked at them—the quiet observer, the naturalist, the journalist—and for the first time, he didn't feel like he was doing science for the sake of science. He was doing it for the spring.

"We're going to need a projector," Declan said, his tone shifting into a tactical plan. "And a screen. We aren't just telling them what's happening. We're showing them. We're going to walk them through the hydrology of the destruction, from the 1974 permits to the 2024 water levels."

Avery nodded. She stood up and walked to the window, watching the moonlight ripple on the surface of the Glasswater. It was lower than yesterday, the water's edge receding into the shadows, but she felt a strange, fierce pride. They weren't watching it die anymore. They were fighting for the revival.

"Let's go," she said, her voice carrying the strength of the currents she knew so well.

They moved out of the shed as a unit, the heavy, humid air of the Florida night closing in around them. They were no longer just friends in a park; they were a team with a target. And for the first time in a long time, the spring felt less like a vanishing memory and more like a challenge that was finally being answered.

The walk through the park was silent, the darkness magnifying the sounds of the night—the rustle of palmettos, the distant hum of traffic on the highway, the rhythmic drip-drip-drip of water falling from an overhanging fern into the shrinking basin. Every sound felt like an accusation.

They reached the perimeter of the Enclave, where the manicured landscaping gave way to the harsh, gray concrete of the drainage system. It was an ugly, industrial scar on the landscape, a concrete culvert that channeled rainwater into a massive, stagnant detention pond.

Micah waded into the tall grass at the edge of the pond, his kit slung over his shoulder. He moved with a practiced, fluid grace, his eyes scanning the murky water. He stopped at a point where the water flowed into the culvert, pulling out a test tube and a handheld colorimeter.

"The runoff is active," he whispered, holding the tube up to the beam of his flashlight. The water inside was cloudy, a murky, brownish-green that shimmered with an oily sheen. "Even with no rain today, there's a constant trickle. Sprinklers. They're running them around the clock to keep the lawns green in the middle of a drought."

Avery watched, her heart sinking. "They're literally pumping the aquifer dry to keep their grass alive, and then polluting the runoff with the fertilizer they use to make it grow."

"It's a closed-loop disaster," Declan agreed, hovering nearby with his drone controller. "If I fly the thermal camera, I can map the exact path of the underground seepage. We can show how this water isn't hitting the water table correctly—it's hitting the limestone shelf and leaching out directly into the Grotto."

He launched the drone, its red and green navigation lights blinking like a mechanical insect against the dark sky. The feed on his tablet lit up with a ghostly, heat-mapped view of the earth.

"There," Declan pointed to a series of glowing blue veins snaking beneath the surface of the Enclave's golf course. "That's the conduit. See how it bypasses the natural filtration zones? It's a direct line to the spring."

Avery leaned in. The truth was starkly, coldly visible on the screen. The blue lines represented the path of the water, a stark contrast to the surrounding dark soil. It was like seeing the veins of a living creature that had been surgically rerouted to feed an artificial parasite.

"They didn't just build on top of the aquifer," Avery said, her voice trembling. "They turned it into their own personal plumbing system."

Zara scribbled notes, her pen scratching against the paper with a sound like a nervous heartbeat. "When we present this at the meeting, they won't be able to claim it's a mystery. They'll have to account for every single one of those pipes."

As the drone circled back, descending into Declan's waiting hands, the group stood in the dark, the reality of their discovery settling in. They were four kids against a decades-old machine, armed with nothing but data and the truth.

But as Avery looked back toward the spring—the water still and dark, the moonlight reflecting off the surface—she realized that the truth had a weight of its own. It was a pressure that, if applied correctly, could move mountains. Or at least, it could force a town to look at the ground beneath its feet and finally see what it was losing.

"We have everything we need," Avery said, her voice clear and calm. "Let's go home and prepare. The Town Hall starts at nine."

The walk back felt shorter, more purposeful. Every step was a piece of the solution, every observation a building block of the case. They reached the shed as the first hint of gray light touched the horizon, the dawn beginning to wake the Florida woods.

They spent the next few hours in a fever of activity—organizing the archival documents, labeling the test tubes, and refining the slides for the presentation. They didn't speak much; they didn't need to. The data spoke for itself.

When the sun finally rose, casting a golden light through the slats of the shed, they looked at each other—tired, disheveled, and entirely ready.

"We're going to change things," Micah said, his hand resting on the stack of documents.

"We're going to fix things," Avery corrected him.

They walked out of the shed into the humid morning air, the sun's light catching the dew on the leaves and the first stirring of life in the park. For the first time, the spring didn't look like it was drying up. It looked like it was waiting.

And as they headed toward the community center, the town of Glasswater didn't look like a sleepy, oblivious community anymore. It looked like a puzzle, one they were finally equipped to solve.

The Town Hall meeting room was already packed. It was a cramped, beige-walled space that usually hosted bingo nights and HOA budget reviews, but tonight the air was thick with a different kind of tension. Old-timers in faded fishing hats sat next to young professionals from the Enclave, the divide between them as palpable as the drought itself.

Avery felt a moment of dizziness as she took her seat at the front. The room was a sea of skeptical faces—people who had spent decades taking the spring for granted, and people who had invested their lives in the development that was destroying it.

Zara stood at the podium, her hands steady, her eyes scanning the room. She didn't wait for the moderator to invite her. She started, her voice amplified and echoing against the acoustic tiles.

"We were told the spring was drying up because of a drought," she began, and the room grew quiet. "We were told that the environment is a cycle, and that eventually, the rain would return. But we investigated. We didn't look at the sky; we looked at the history."

She clicked a remote, and the screen behind her flashed to life. The 1954 map appeared, the red ink standing out against the faded, grainy background.

"In 1954, this land was a primary recharge zone for the Floridan Aquifer," Zara continued, her voice clear and authoritative. "It was the heart of our community's water supply. But then, it was rezoned. It was 'disturbed.' It was paved."

She clicked again. The modern, glossy overlay of the Enclave appeared, the grid of concrete blocks stark against the natural topography. "This isn't a natural drought. This is an engineering failure. We've paved over the lungs of our spring, and we're surprised it can't breathe."

The murmurs began—first a trickle, then a wave. People shifted in their seats, looking at each other, their faces a mix of confusion, anger, and mounting realization.

Declan stepped up next, the drone's thermal maps projected on the wall behind him. "This is a heat map of the underground drainage system," he explained, his voice the calm, practiced tone of a scientist. "These blue lines are the conduit. This is where the rainwater goes instead of recharging the spring. It's being funneled into the sewer system, carrying fertilizers and pesticides directly into the Grotto."

He pulled up the chemical reports Micah had collected. "These are the nitrate levels from the runoff. They are ten times the EPA standard for safe water. This isn't just about the spring running dry. It's about the fact that the very maintenance of our own homes is poisoning the water we're drinking and the water that supports our ecosystem."

Micah took the stage then, his voice quiet but intense. "We don't need a miracle to bring the spring back. We need to restore the path of the water. We need to dismantle the drainage systems that are cutting off the recharge, and we need to return the runoff to the land where it belongs. We have the technology, and we have the capacity. All we need is the will."

Avery stood up and walked to the front. She didn't use the microphone. She let the silence of the room draw the audience in. She thought about the spring—the way the water felt on her skin, the way the sunlight danced through the glass-clear pool, the way the fish clustered in the dark.

"The spring isn't just a resource," she said, her voice soft but carrying to the back of the room. "It's a witness. It's been here for thousands of years, watching us build, watching us grow, watching us make choices. And now, it's showing us the consequence of those choices. We can either keep pretending that we don't know what's happening, or we can choose to be the generation that actually listens to the land."

She looked out at the room. She saw the mayor, his face pale and sweaty. She saw the developer, his eyes narrowed, staring at the screen. But she also saw the people—the families who came to the spring on weekends, the retirees who had lived here for forty years, the neighbors who cared about the place they called home.

The silence in the room wasn't the silence of skepticism anymore. It was the silence of a community that was being forced to face its own reflection.

"We have a plan," Avery concluded, gesturing to the stack of reports and maps on the table. "We have the data, we have the solutions, and we have the historical context to make sure this never happens again. The question is, do you have the courage to fix what we've broken?"

For a moment, there was no sound at all. Then, the back of the room erupted.

It wasn't a roar of approval, exactly—it was the noise of a hundred conversations starting at once. It was the sound of a community finally starting to argue about the right things.

Zara leaned into the microphone, her eyes bright with a mixture of exhaustion and triumph. "We're going to be here all night," she said, her voice projecting over the chaos. "We're going to talk about the permits, the zoning, and the restoration project. And we're not leaving until we have a vote."

Avery sat back down, her legs shaking, her heart hammering against her ribs. She looked at her team—Declan, already surrounded by a circle of curious homeowners; Micah, explaining the nitrate levels to an earnest-looking teacher; and Zara, taking notes as fast as she could.

They had done it. They hadn't fixed the spring, not yet—that would take years of work and gallons of sweat. But they had broken the silence. They had moved the needle from 'misfortune' to 'malpractice,' and from 'drought' to 'development.'

As the meeting hummed along, the energy in the room shifting from defensive to collaborative, Avery looked up at the clock. It was late, but for the first time, she felt like she had all the time in the world. The spring was waiting, and now, so was the town.

She leaned back, a small smile playing on her lips. The Glasswater was still there, beneath the surface, beneath the concrete, beneath the bureaucracy. It had survived everything else. And now, it finally had an army.

The vote was unanimous, a rare, startling consensus that left even the most cynical council members blinking in surprise. By midnight, the proposal for the "Glasswater Recharge Initiative" had been ratified. It was a commitment to tear up the most critical sections of the Enclave's paving, to restore the natural swales, and to install bio-retention basins that would filter the runoff before it could hit the limestone.

It wasn't a quick fix, and it certainly wouldn't be easy. There would be lawsuits, construction delays, and months of political maneuvering. But the foundation had been laid. The community had accepted their role in the damage, and they had committed to being part of the solution.

As the crowd filtered out of the hall, the cool night air hitting their faces, the sense of accomplishment felt almost heavy, like a physical weight.

Avery walked out into the parking lot, the others trailing behind her. The moon was high, casting the trees in the park into sharp, dark relief. The distant, faint roar of the spring—the sound of the water moving through the earth—seemed just a little bit clearer, a little bit stronger.

"We did it," Declan said, his voice husky with fatigue. He slumped against his car, his drone case tucked under his arm. "They actually voted for it."

"They had to," Zara replied, checking her tablet one last time. "The optics were too bad. And the engineering, the data... it was airtight."

"It wasn't just the data," Micah said, looking toward the park. "It was the truth. People just needed someone to point it out to them."

Avery looked at the park, her heart swelling. "It's going to take time. The water won't be back by morning."

"No," Declan agreed, looking at the sky. "But it's a start. We've given the aquifer a chance to heal."

They walked to the edge of the parking lot, where the light from the building faded into the shadows of the woods. They were exhausted, drained, and changed in ways they couldn't yet articulate. They were the ones who had unraveled the mystery, but they were also the ones who had to live with the responsibility of what they had discovered.

"What now?" Zara asked, her voice quiet in the stillness.

Avery turned to them, her face illuminated by the soft light of the stars. "Now? Now we monitor. We make sure they keep their word. We watch the flow rates. We document the vegetation. We make sure the 'remediation' actually happens."

She looked at her team—the scientist, the journalist, the naturalist. The ones who had stood by her when everything seemed hopeless.

"We're not done," she said, her voice steady and sure. "We're just getting started."

They walked back to their cars in silence, the weight of the night settling on them. But as they drove away, Avery watched the park in her rearview mirror. The trees, the darkness, the spring—they were there, enduring, waiting for the water to return.

The next morning, the air was different. The humidity had returned, a heavy, familiar blanket, but it carried the faint, sweet scent of rain.

Avery walked down the boardwalk to the Grotto, her gear packed and ready. She reached the edge of the platform and looked down.

The water was there.

It was still low, but the flow had shifted. It was no longer stagnant. The clear, crystalline water was bubbling up from the vent, a small, tentative pulse of life that seemed to be pushing against the limestone shelf.

Avery gasped, her hand going to her throat. She knelt on the edge of the boardwalk, leaning over the water. She watched as a small school of minnows darted into the center of the boil, catching the current.

It wasn't a roar. It wasn't the powerful surge that had defined the spring for centuries. But it was there.

She took a picture, the shutter clicking softly in the quiet. It was the first of many she would take in the months to come—a record of the recovery, a timeline of the healing.

She felt a hand on her shoulder. Declan stood beside her, his face lit with pure, unadulterated wonder.

"The hydrostatic pressure is up," he whispered, checking his device. "Only by a fraction, but it's real. The recharge is already beginning to influence the flow."

Micah joined them, looking down at the water. "The plants at the edge are already perking up," he noted, pointing to a cluster of maidenhair ferns that had looked dead a few days ago. "They know. The ecosystem knows the change is coming."

Zara walked up behind them, her notebook held like a treasure. She leaned in, looking at the water, then at her team. "What are you going to write for the update?" she asked, a small smile on her lips.

Avery looked at the water—the clear, cold, living heart of the spring. She thought about the map, the data, the meeting, the fight. She thought about the way the community had finally looked at the land and seen it for what it was.

"I don't think I need to write anything," Avery said, her voice quiet. "I think the spring is going to tell the story for us."

As they stood there, watching the water rise, the first real rain of the season began to fall. It started as a light, misty drizzle, then intensified into a steady, soaking downpour.

The water in the spring didn't just rise; it seemed to shimmer, the surface tension broken by the thousands of tiny raindrops. It was the sound of the world finally exhaling.

Avery smiled, the rain soaking into her clothes, her hair, her skin. It was the most beautiful sound she had ever heard. The spring was alive. And for the first time, she knew that it was going to stay that way.

She reached into her bag, pulled out her camera, and set the aperture. She waited for the right moment, for the light to hit the water just so, and then she clicked.

The photo showed the water, the rain, the life. But more than that, it showed the beginning. It was the first image of a long, difficult, beautiful recovery. And as she watched the rain fall, she knew that she would be there to capture every single step of the way.

The Glasswater was flowing, and the story had only just begun.

Chapter Fifteen

CONFLICT ON THE BLUFF

The morning sun hammered against the asphalt of the Enclave's visitor center, reflecting a blinding, artificial glare that made the lush green of the nearby preserve look like a faded photograph. Zara adjusted her recording device, the slight tremor in her hand hidden by the professional grip she kept on her notepad. Beside her, Micah stood unnaturally still, his boots caked with the red clay he'd been scraping from the storm-drain grates all morning.

Mr. Henderson, the Enclave's site manager, leaned back in a chair that cost more than a semester of community college. He kept his sunglasses on, rendering his face a mirrored wall.

"I'm going to stop you right there, Ms. Patel," Henderson said, his voice smooth, practiced, and utterly dismissive. "You're what? Seventeen? Eighteen? You're asking for access to our private irrigation schematics because you saw a couple of dry spots in the basin. Do you have any idea how many permits we cleared before we broke ground? We aren't just developers; we're partners in this town's growth."

"Growth isn't the problem, Mr. Henderson," Zara said, stepping forward. Her voice was steady, sharpened by the hours she'd spent scouring public records. "The drainage rerouting is. We've mapped the surface flow from the back nine of your development. It bypasses the natural limestone sinkhole near the ridge—the one that feeds the main aquifer—and dumps directly into the retention pond. You've essentially choked off a lung."

Henderson chuckled, a dry, rattling sound. "A 'lung.' Very poetic. Did you get that from a textbook, or did you pull it out of the swamp?"

Micah's jaw tightened. He took a half-step forward, his voice low, vibrating with a restrained intensity that was a far cry from his usual gentle demeanor. "The vegetation along the bluff is dying, sir. Not just the decorative stuff you planted, but the scrub oaks and the native ferns. They're thirsty because the water table has dropped six feet in two weeks. That isn't poetry. It's a collapse."

Henderson stood up, his chair clattering back. He walked to the window, gesturing grandly toward the neatly manicured grass of the development. "Look at this. Green. Lush. Sustainable. We pay for the landscaping, we pay for the water, and we contribute to the tax base that keeps your little park rangers in their uniforms. Are you coming in here with amateur-hour sketches and trying to blame us for a regional drought? It's Florida, kids. It rains, it dries. That's the cycle."

"It's not a drought cycle," Zara pressed, ignoring the condescension. "The flow rate at the spring is down forty percent while your irrigation usage has tripled. The correlation isn't 'amateur-hour,' it's mathematical."

"It's harassment," Henderson countered, his tone hardening. "If you set foot on our property line again, or if you continue to solicit our contractors for 'data,' I'll have security escort you off the bluff. And trust me, the town council doesn't appreciate kids disrupting the local economy with conspiracy theories."

Micah stepped between Zara and the developer, his hands open—a natural gesture of peace that looked, in this light, like a man trying to stop a flood with his palms. "Mr. Henderson, if the spring dies, the water quality in your own retention ponds will degrade. The salinity will rise. You're killing your own investment. We're offering a chance to fix the diversion before it becomes an irreversible catastrophe."

Henderson didn't even look at him. He checked his watch. "The conversation is over. Get out."

The walk back to the truck was heavy with the humid, oppressive weight of the Florida afternoon. As they reached the edge of the development, the silence between them wasn't the companionable quiet of a research trek; it was jagged and frayed.

Zara slammed the truck door, the sound echoing against the dense pines. She stared out at the shimmering heat haze rising from the pavement. "He didn't even blink. He knew exactly what we were talking about, and he didn't care."

Micah sat in the driver's seat, his hands gripping the steering wheel so hard his knuckles turned white. He exhaled a long, shuddering breath, trying to force his heart rate back down. "He thinks we're a joke. He's looking at the ledger, not the land."

"We *are* a joke to them, Micah," Zara snapped, her frustration finally boiling over. "We're four kids with a drone and a couple of notebooks. We have no authority. We have no institutional backing. We went in there and handed him our entire hand of cards, and he just laughed."

"We had to try," Micah murmured, though the doubt was clearly gnawing at him. "I thought if he saw the impact on the local ecosystem—"

"He doesn't care about the ecosystem!" Zara cut him off. "He cares about liability. And until we can prove that his actions are a direct, legal liability to the town, he's going to keep the water diverted."

She leaned her head against the cool glass of the window, her mind racing. "We've been acting like this is a journalism project or a nature walk. It's not. It's a deposition. Every observation we've made—Avery's photos, Declan's sensor readings, your vegetation maps—it's all anecdotal until we tie it into a single, unbreakable narrative. We need to stop chasing 'stories' and start building a case."

Micah looked at her, his eyes searching hers for the spark of optimism he usually relied on, but finding only cold, hard resolve. "You're talking about a level of scientific rigor that we aren't equipped for. We aren't hydrologists, Zara. We're students."

"Then we learn," she said. "We stop relying on what we *feel* and start focusing on the raw, ugly, irrefutable data. We need to prove the water movement, track the nitrates, and map the exact conduit under the limestone. No more 'I think the spring looks lower.' We need to be able to tell the council exactly how many cubic feet of water are being stolen every hour."

"And if we find out we're wrong?" Micah asked, his voice barely a whisper. "What if the data says it's something else?"

Zara looked at the spring through the trees, a faint, shimmering ribbon of blue fighting for its life against the encroaching brown of the parched banks. "Then we follow the data wherever it leads. But we aren't going back to Henderson until we can prove he's guilty."

She pulled out her phone and started typing, her thumbs flying across the screen.

Emergency meeting at the Grotto. Bring everything—no more anecdotes. We're shifting the strategy.

Micah watched her for a moment, the tension in his shoulders slowly dropping, replaced by a grim, focused determination. He started the truck, the engine turning over with a low, steady growl. He didn't look at the Enclave's gate as he pulled away; he kept his eyes on the road, watching the way the shadows fell across the forest floor, marking the places where the earth was already starting to fail.

The investigative mission had shifted. It was no longer about discovery—it was about defense. And for the first time, they were no longer just observers of the spring; they were its legal and scientific shield.

The heat inside the truck was suffocating, a heavy, humid blanket that smelled of stale pine needles and the lingering ozone of Micah's fraying temper. Zara gripped the steering wheel so tightly her knuckles were white, staring straight ahead at the dirt road that wound away from the Enclave's manicured perimeter.

"They aren't just dismissive," Micah said from the passenger seat, his voice vibrating with a restraint that felt like a coiled spring. He looked out the side mirror at the receding gated entrance. "They're dangerous. They have an entire infrastructure built on the assumption that the spring is a bottomless resource, or that it's simply... irrelevant."

In the back, Declan was hunched over his laptop, his face illuminated by the harsh blue glow of sonar topographical maps. Avery sat beside him, watching the data points flicker. She was quiet, her hands resting on the strap of her camera, her thumb tracing the groove of the shutter button. The silence in the back was different—heavy, expectant, and brittle.

"They didn't even look at the drone footage," Declan muttered, his eyes never leaving the screen. "I showed them the thermal signature of the runoff bypass. I showed them, in high definition, exactly where their storm drains are cutting into the limestone shelf. They just laughed. Henderson asked me if I wanted a job in PR because I had such an 'active imagination.'"

"Because that's how you kill a story, Declan," Zara snapped, though her voice lacked its usual sharp edge. She slowed the truck as they hit a patch of rutted gravel, the suspension groaning in protest. "You discredit the source. If they make us look like kids playing scientist in the woods, they don't have to answer to the board. They don't have to answer to the town."

"Maybe we are," Avery said, her voice soft but cutting through the tension like a stone through water. Everyone looked at her. "Maybe we're just kids playing scientist. Look at what we have. A few journals, some thermal pictures, and a bunch of hunches. If we go to the Town Hall meeting with this, we're going to be eaten alive."

Micah turned fully in his seat, his brow furrowed. "Avery, the spring is dying. We've seen the vegetation thinning. The flow rate at the headspring is down twenty percent from last year. We aren't 'playing.'"

"I know that," Avery replied, her gaze fixed on the gear shifting on the center console. "But *knowing* it and proving it to people who have the money to bury us are two different things. Henderson is right about one thing: we're fragmented. You're looking at nitrates, Micah. Declan is looking at conduits. Zara is looking at property deeds. We're four separate investigations, not a unified front. Every time they ask us a question, we stumble because we're only looking at our own slice of the pie."

"She's right," Declan admitted, a rare moment of vulnerability crossing his face. He shut his laptop screen halfway. "My sonar data is compelling, but without the chemical composition data to prove that the water in those conduits is actually tainted, it's just a map of a hole in the ground. It could be an empty pipe, for all a lawyer would care."

Zara pulled the truck onto the shoulder of a logging road, cutting the engine. The sudden silence of the Florida scrubland, punctuated only by the rhythmic clicking of the cooling engine and the distant, lonely call of a limpkin, felt absolute. She turned in her seat, facing them.

"So, what's the alternative?" Zara asked. "We stop? We let them pave over the rest of the recharge zone while we spend three months trying to get a PhD in hydrology?"

"No," Avery said, her confidence growing. She reached into her bag and pulled out a battered field notebook, thick with pressed leaves and scribbled notations. She laid it across the center console. "We stop acting like independent researchers. We need to build a single, unbreakable timeline. Declan, I need the exact GPS coordinates of every limestone karst you've mapped. Micah, I need your water samples for the last six weeks, cross-referenced with every rain event on the local weather logs. Zara, stop digging for stories and start digging for connections. Find out who signed off on the 1974 drainage permits. I want to see the names. I want to see the paper trail that allowed the Enclave to exist on top of our water table."

Micah leaned forward, watching as Avery tapped the page with her pencil. "You want us to synchronize."

"I want us to treat this like a surgical operation," Avery said. "No more 'maybe' or 'it seems like.' If we're going to challenge them, we need a report that looks like it came from an environmental engineering firm. If we can show that the Enclave is not just hurting the spring, but violating their own site-plan stipulations from the seventies, they can't call us alarmists. They'll have to call us a liability."

Declan started tapping on his laptop again, his movements sharper, more deliberate. "I can overlay the nitrate spikes from your samples directly onto the thermal runoff maps, Micah. If the spikes happen within six hours of a localized rainfall, and the thermal data shows the runoff hitting the aquifer at those exact points... that's not a coincidence. That's causation."

"And if I find the original site-plan covenants," Zara added, her journalistic fire reigniting, "I can show that they've effectively modified the landscape in direct violation of the original environmental impact mitigation strategy. It's not just a moral issue anymore. It's a breach of contract."

The air in the truck shifted. The frustration that had fueled their heated standoff with the developers was distilling into something colder, harder, and infinitely more focused.

Micah looked at his hands, then out at the dense wall of saw palmettos. "I've been approaching this with my heart," he said quietly. "I wanted to show them the beauty of the spring so they'd care. But you're right, Avery. They don't care about the beauty. We have to speak their language."

"Data," Declan said, his eyes reflecting the screen. "We're going to drown them in data."

Avery pulled her camera up, checking the lens. "We have the rest of the weekend. We consolidate every piece of evidence. No more solo missions. We hit the site at dawn, we map the final points, and we don't talk to a single soul until we have the full picture ready for the board."

Zara reached out, placing her hand on the center console. One by one, the others placed their hands over hers—a quiet, unspoken pact in the heart of the humid wilderness. The girl who watched the currents, the boy who mapped the stone, the woman who hunted the truth, and the man who stood for the lifeblood of the forest. They were no longer just four students on a weekend exploration. They were a collective, and the silence of the woods seemed to acknowledge the shift.

"Okay," Zara said, her voice steady. "Let's get to work."

Declan fired up the drone, the whine of the rotors cutting through the heavy air, rising above the canopy. Avery adjusted her settings, ready to capture the evidence that would force a reckoning. As the drone drifted upward, the spring—hidden behind the curtain of cypress and fern—remained silent, waiting for the defense they were now armed to provide.

Chapter Sixteen

THE SINKHOLE CONNECTION

The northern sector of the park didn't officially exist on the visitor maps. It was a jagged expanse of limestone scrub and dense, strangling palmetto thickets that had been cordoned off years ago, ostensibly for "ecological restoration."

Avery wiped sweat from her eyes, the humidity thick enough to swallow a scream. Ahead of her, Declan hacked at a wall of sawgrass with a machete he'd borrowed from the maintenance shed. He wasn't just walking; he was hunting.

"The drone feed was clear, Avery," Declan said, his voice strained. He stepped over a fallen cypress log, his boots sinking into the dark, spongy muck. "Thirty yards past this ridge. The ground elevation drops three feet in a span of ten yards. That's not natural settling. That's a depression trap."

Avery kept her eyes on the ground. She wasn't looking at the horizon; she was looking at the soil. The moss here was grey, brittle—not the lush, vibrant green of the basin near the main spring. It was dying. The silence in this part of the woods was heavy, missing the usual chorus of insects and birds.

"Wait," Avery said, grabbing his arm.

Declan froze, the machete held mid-air. "What?"

"Look at the roots." She knelt, pushing aside a layer of decaying pine needles. The base of an ancient slash pine was exposed. The soil beneath it was hard, packed down with a grey, cement-like slurry. "This isn't forest floor. It's construction debris. Someone tried to bury a depression."

Declan knelt beside her, his face turning pale under his freckles. He pulled a folding spade from his pack and dug a shallow trench. His blade hit something solid with a metallic *clink*.

"Concrete," he muttered. "Reinforced. This is a sinkhole throat. It's the primary recharge point for the northern aquifer, and they've capped it like a drain in a bathtub."

"They didn't just cap it," Avery whispered, the realization sending a chill through her despite the oppressive heat. "They leveled the surrounding land to force runoff toward the Enclave's retention ponds, then filled this to stop the recharge. They choked the spring at its source."

The trek became a grim processional. As they pushed deeper into the restricted zone, the landscape became increasingly jagged. They found illegal dumping—piles of crushed limestone, rusted rebar, and bags of uncured concrete shoved into the crevices of the limestone shelf. It was systematic. It wasn't an accident; it was an industrial-scale rerouting of the water table.

Declan pulled a handheld GPS unit from his pocket, his hands shaking slightly. "The drone footage showed a series of these. They're all connected by an artificial trench hidden under that canopy of vines. Avery, if these were open, the spring would be at capacity, even in a drought."

They reached the anomaly Declan had tracked. It was a massive, unnatural crater, roughly twenty feet across, filled with discarded construction pallets and heavy, sodden clay. A trickle of stagnant water sat at the bottom, oily and iridescent.

"This is it," Declan whispered. "The primary artery. They plugged the heart of the spring to keep their own drainage ponds from overflowing."

Avery took a photo, the flash cutting through the dim, filtered light of the scrub. She felt a surge of cold fury. This wasn't just development; it was a crime against the very land that sustained them. She thought of the spring, of the way the water used to shimmer—the living, breathing blue of her childhood. It was being suffocated to keep a manicured lawn green a few miles away.

"We have to move," Avery said, her voice steadying. "If we stay here, we're trespassing. If we get caught, they'll bury the evidence along with this sinkhole."

Declan stood, holstering his gear. His adventurous swagger was gone, replaced by a cold, hardened determination. "I have enough data. I have the geolocations, the depth profiles, and the chemical markers from the soil samples I took back there. This isn't just a theory anymore, Avery. This is a map of a crime scene."

As they began the long trek back toward the main trail, the sky above the canopy began to rumble. A Florida storm was brewing, the clouds darkening to a bruised purple. The humidity spiked, the air turning metallic.

"We're going to get caught in it," Declan noted, his eyes scanning the dense brush.

"Let it rain," Avery said, pushing through a thicket of scrub oak. "If the water starts rising, we'll see exactly where it's *not* going. We'll see the blockade in action."

They moved with frantic urgency, the terrain becoming more treacherous as the first fat, heavy drops of rain began to slam against the leaves. The mud underfoot turned to slick, treacherous grease. Every step was a battle against gravity and the encroaching forest.

When they finally broke through the tree line and onto the main service road, the sky opened up. A torrential downpour turned the road into a river, but it wasn't flowing toward the spring. They watched, breathless and soaking wet, as the runoff surged in the opposite direction, channeled away by a series of graded embankments that Declan had pointed out on his drone maps.

"Look at it," Declan shouted over the roar of the wind, pointing to the water diversion. "It's bypassing the basin entirely. They built a bypass, Avery. They're literally throwing the spring's lifeblood into the drainage ditch."

Avery didn't answer. She watched the water move, her mind tracing the flow, connecting the craters they'd found to the drought-stricken spring. The puzzle was no longer missing pieces. She had the entire shape of the betrayal in her head, burned into her memory as clearly as the storm lashing at their faces.

"Let's go," she said, pulling her bag tight against her chest to protect the camera. "We need to get this to the others. Now."

They sprinted back toward the research hub, their legs burning, their hearts hammering against their ribs. The storm raged behind them, but the internal storm was far more powerful. They weren't just two teens playing at investigation anymore. They were witnesses to a systematic dismantling of their home, and for the first time, they held the evidence to tear that system down.

As they reached the edge of the park's main clearing, the lights of their research hub glowed like a beacon in the rain. Micah and Zara were inside, unaware that the map they were about to build would change everything.

Avery paused for a heartbeat at the door, catching her reflection in the dark, wet glass. She looked fierce, her hair matted with leaves, her clothes ruined, but her eyes were clear. The quiet observer was gone. In her place was someone who understood exactly what the land needed, and what it would take to fight for it.

She pushed the door open, the bell above it chiming in the rain-heavy air.

"We found it," she said, her voice cutting through the hum of the equipment.

Declan stepped in behind her, dropping his heavy, mud-caked pack onto the table with a final, resonant thud. The sound echoed in the small room, grounding them, marking the end of their search and the beginning of the war.

The humidity inside the research shed was stifling, a thick, pressurized blanket that smelled of damp limestone and ozone. Avery wiped a smudge of red clay from her forehead, her hands trembling slightly as she set the sample vials down on the workbench. Beside her, Declan was hunched over the main monitor, the blue light of the software reflecting in the lenses of his glasses, making him look less like a high school senior and more like a weary forensic engineer.

"Look at the topology, Avery," Declan said, his voice tight. He tapped a key, and the digital map of the Northern Sector flickered, overlaying his drone's thermal data with the historical land-use surveys Zara had unearthed. "It's not just a few clogged sinkholes. It's a systemic chokehold."

Avery leaned in, squinting at the screen. The map was a web of contradictions. The natural karst topography—the veins of the earth—was clearly marked in soft, flowing lines. But the modern data, the layers they had meticulously pieced together over the last seventy-two hours, showed something jagged and brutal: a series of engineered concrete baffles and illegal fill zones.

"They didn't just pave over the recharge zones," Avery whispered, tracing the path of the blocked conduit on the screen. "They redirected the entire sheet flow of the northern bluff into the Enclave's retention ponds. They've been starving the spring to keep their lawns emerald green."

"It's a massive hydraulic bypass," Declan agreed, his fingers flying across the keyboard. "When we were out there, I thought it was just poor land management. But this? This is calculated. Someone knew exactly where the primary limestone conduits were. They diverted the water at the source, right before it could reach the swallow holes. They effectively put a tourniquet on the spring's throat."

The cabin door creaked open, and Micah stepped inside, his boots caked in the same distinctive red mud that clung to Avery's jeans. He was carrying a cluster of wilting ferns, his face etched with a quiet, devastating anger.

"The flora along the riparian corridor is failing," Micah said, his voice dropping into that somber, reverent tone he used when speaking about the forest. "It's not just water volume, guys. The runoff from the Enclave... It's saturated with high-nitrogen fertilizers. I've run the tests on the standing water near the drainage spillways. The nitrate levels are ten times higher than the baseline for this ecosystem. It's toxic. It's killing the native filter-feeders that keep the spring basin clear."

Avery felt a surge of cold clarity. "So, they're blocking the clean water and pumping in the contaminated runoff."

"It's a double death sentence," Declan said, finally pulling away from the monitor and turning to face them. "If we just cleared the sinkholes, the spring would be flooded with that poisoned runoff within a week. The restoration has to be holistic, or it's useless."

Avery pulled a chair over, dragging it across the wooden floor with a screech that made them all jump. She reached into her gear bag and pulled out the physical maps they had marked up—crinkled, stained with swamp water, and covered in their scribbled notations. She spread them out over the digital tablet, creating a literal and figurative bridge between their evidence.

"Zara's in the archives now," Avery said, her voice finding a new, steady rhythm. "She's got the permit records from 1974. If she can link these specific concrete fills to the original development covenant—the one that mandated the preservation of natural drainage—we don't just have an environmental argument. We have a legal one."

"It's a smoking gun," Micah said, leaning over the table. He traced the intersection of the Enclave's drainage pipes and the natural underground flow paths Declan had mapped. "We've been acting like this was a mystery to solve. It's not a mystery. It's a liability."

Declan looked at the map, then up at his friends. The uncertainty that had haunted their early days—the fear that they were just teenagers chasing ghosts—had evaporated. In its place was a hard, sharp edge of conviction.

"We need to synthesize this," Declan said, pulling up the master file. "If we show the community the direct line between their property values and the death of the spring, it won't be 'teenager alarmism' anymore. It'll be a structural assessment of their own infrastructure failure."

Avery grabbed a marker, stepping up to the large whiteboard on the back wall. She began to draw the arc of the spring's flow, adding the symbols Declan had provided for the conduits, then adding Micah's data points for the toxicity zones. As she worked, the individual fragments—the dry sand, the wilting ferns, the buried cave systems, the bureaucratic lies—began to coalesce into a terrifying, undeniable picture of human interference.

"We have to make them see it," Avery said, capping the marker with a sharp snap. "Not just hear it. They need to see the map."

"They'll try to shut us down," Micah cautioned, though his eyes remained fixed on the whiteboard. "The homeowners' association at the Enclave... they're powerful. They have the ear of the council."

"Then we'll make it impossible for the council to look away," Declan said, his voice devoid of his usual geological abstraction. He looked at

Avery. "You've got the photos of the spring's decline—the loss of the manatee grass, the thinning of the flow. I've got the CAD models of the subterranean blockage. Micah, you've got the chemistry that proves the water is being poisoned."

"And I have the paper trail that proves they never had the right to build that drainage system in the first place," Zara's voice cut through the air.

The door swung open fully, and Zara walked in, her face flushed with excitement, clutching a thick stack of photocopies. She looked like she hadn't slept in two days, her hair a wild tangle, but her eyes were glowing with the predatory focus of a journalist who had finally found the thread that would unravel the sweater.

"I found the minutes from the '74 zoning board meeting," Zara said, dropping the papers onto the workbench with a heavy thud. "They explicitly warned the original developers that the northern limestone shelf was a 'critical recharge zone.' It was a protected geological feature. They built the drainage infrastructure in direct violation of the environmental impact mitigation statement."

She looked at the whiteboard, then at the team. The silence in the room was heavy, not with tension, but with the sudden, gravity-shifting weight of their combined discovery.

"It's not just a mess," Zara added, her voice dropping to a whisper. "It's a crime."

Avery felt a chill run down her spine. The spring, the silent, ancient heartbeat she had listened to her entire life, had been fighting a war she hadn't even known was happening. She looked at the map on the board—the red lines of the runoff, the black ink of the barriers, the blue flow of the spring that was struggling to push through the debris.

"We have everything we need," Avery said, her voice firm. "We're not just kids talking about a drought anymore. We're holding the receipts for the destruction of the Glasswater aquifer."

Declan nodded, his pragmatic edge returning. "The Town Hall is in two days. That's enough time to refine the visuals, map out the timeline, and get the legal citations organized."

"And then?" Micah asked, looking at the map. "What happens when we walk into that room?"

"Then," Avery said, turning back to the spring, which she could see through the shed's single, dirt-streaked window, "we stop asking for permission to save it. We start telling them how we're going to fix it."

The team stood in the dim light of the research shed, a ragtag group of students surrounded by the tools of their discovery. The air felt thinner, sharper, as if the oxygen in the room was being pulled toward the weight of their decision. For the first time, the path forward wasn't buried in the cave systems or hidden in the archives. It was clear, direct, and entirely theirs to navigate.

Avery reached out and touched the map on the board, her fingertip resting over the spot where the main conduit had been choked with concrete. She could almost feel the phantom pressure of the water beneath her skin—the push, the urgency, the slow, rhythmic pulse of the spring waiting for someone to clear the way.

"Let's get to work," she said.

Chapter Seventeen

Roots of the Runoff

The dawn light filtered through the canopy in thin, golden needles, illuminating the fine mist rising from Glasswater. It should have been a morning of celebration, but the air held a heavy, investigative stillness. Avery adjusted the strap of her camera, her boots sinking into the soft, damp mulch near the northern perimeter of the park.

Micah followed a few paces behind, his eyes fixed on the vegetation line where the lush ferns gave way to a suspicious, brownish scrub. He carried a small, portable water-testing kit, a relic he'd borrowed from the university lab.

"Look at the *Salvinia*," Micah murmured, pointing toward a stagnant patch near a drainage culvert. "It's thriving in places it shouldn't be. That's not natural growth, Avery. That's a symptom."

Avery knelt, bringing her macro lens close to the water's edge. She captured a sequence of shots—the unnaturally vibrant green of the invasive ferns, the film of algae coating the rocks like a neon shroud. "Nitrates," she whispered, the word feeling bitter on her tongue.

"Exactly," Micah said, kneeling beside her. He dipped a test strip into the runoff trickling from the drainage pipe. This pipe ultimately connected to the culverts running beneath the manicured lawns of the suburban housing developments bordering the park. "When people want that perfect, golf-course-ready emerald grass in the middle of a Florida summer, they over-fertilize. The first heavy rain washes it all into the storm drains, and where does it go? Straight into our aquifer."

Avery tracked the pipe's path with her eyes, tracing it mentally beyond the park fence, toward the rows of neatly trimmed hedges and pristine driveways of her own neighbors. She knew these people. She'd bought lemonade from their kids; she'd walked these streets a thousand times without ever looking at the storm drains as anything other than gray concrete scars.

"It's not just the big developers, is it?" Avery asked, her voice quiet.

Micah sighed, his posture drooping. "It's the cumulative effect. A little bit of fertilizer here, an extra ten minutes of automatic sprinkler time there. They think it's harmless because it's out of sight, out of mind. But looking at the biological markers today... the nitrogen levels are spiking. It's starving the spring of oxygen, suffocating the very pulse of the ecosystem."

Avery stood up, her gaze shifting to a cluster of native wildflowers that had been bleached white, their stalks wilted near the discharge point. It wasn't just a technical problem; it was a personal betrayal of the land she loved. She thought of her neighbor, Mrs. Gable, who took such immense pride in her rose bushes. Mrs. Gable was a kind woman, a grandmother who baked cookies for the neighborhood, yet here was the evidence of her 'pride' poisoning the water table.

"They don't know," Avery said, though it felt like a hollow defense. "They really think it's just about having a nice yard."

"Ignorance is a luxury the ecosystem can't afford anymore," Micah replied gently. He stood, gathering his kit. "We aren't just looking at a pipe, Avery. We're looking at the anatomy of our own community's selfishness. It's hard to swallow, but if we don't map this path, we can't fix it."

They walked in silence for a while, mapping the secondary drainage points. Every few hundred yards, they found another culvert. Some were choked with plastic bottles; others, like the one feeding the northern basin, were actively leaching the chemical cocktail of lawn chemicals directly into the recharge zone. Avery snapped photos of every one, her camera acting as a cold, clinical witness.

The weight of it began to settle in her chest—the realization that the "Glasswater Recharge Initiative" wasn't just about moving dirt or rerouting pipes. It was going to require a fundamental shift in how her neighbors perceived their own backyards. She had spent her life observing the spring, thinking it was a separate, wild entity that needed only to be left alone. Now, she understood that the spring was a mirror—a direct, liquid reflection of the community's daily habits.

They circled back toward the central basin. The sunlight was stronger now, catching the surface of the water. Even with the slight recovery they'd seen after the rains, the clarity was still wrong. There was a milky, opalescent quality to the water—the classic sign of an overloaded system.

"You okay?" Micah asked, seeing the way Avery gripped her camera, her knuckles white.

"I'm just thinking about the meeting," she said. "How do we tell them? How do we tell the people who live in those houses that their sprinkler systems are killing the one thing they claim to love about living here?"

"We don't tell them it's 'killing' the spring," Micah suggested, his voice steadying. "We show them the beauty that's being lost. We show them the species that aren't coming back because of what's flowing off their lawns. We invite them to be the stewards of the water, not the ones wasting it. It's a transition from 'blame' to 'partnership.'"

Avery looked at the drainage pipe one last time, a jagged piece of dark plastic protruding from the limestone embankment. She knew the truth now—it wasn't some distant, faceless corporate villain that was the primary threat. It was the collective weight of dozens of small, well-meaning, and entirely disconnected decisions.

"Partnership," she repeated, the word tasting better than 'guilt'. She adjusted her lens, focusing on a single, struggling water lily near the drain. She snapped the photo, framing the flower against the backdrop of the concrete pipe. The juxtaposition was stark, a visual argument that couldn't be ignored.

They began the walk back to the clearing where the truck was parked, the sun climbing higher, turning the humidity into a thick, suffocating blanket. The work was far from over, but as they crested the final hill, the sight of Declan's drone hovering in the distance, a small, buzzing sentinel over the grotto, gave her a strange, sharp sense of purpose. They had the data. They had the reality. Now, they had to prepare for the hardest part: turning the mirror toward the people who needed to see what they had done.

The humidity hung heavy over the spring, a thick, damp blanket that smelled of decaying sawgrass and the metallic tang of drying limestone. Avery stood at the water's edge, her boots sinking into mud that should have been submerged under two feet of crystal-clear, bubbling water. Beside her, Micah adjusted his grip on a water-testing kit, his face etched with a grim, weary focus.

They had spent the better part of the morning tracing the drainage pipes that fed into the spring's basin. The results were not just scientific anomalies; they were personal. The high nitrate spikes they had been tracking didn't originate from some distant, industrial monster. They were flowing directly from the manicured, weed-free lawns of the Highland Estates, just a half-mile upstream.

"It's not just the drought," Micah said, his voice barely audible over the hum of dragonflies. "It's the sheer volume of fertilizer and irrigation overflow. It's like the earth has been turned into a colander, and we've been pouring poison through it for years."

Avery looked toward the boardwalk, where Declan and Zara were waiting. She felt a hollow ache in her chest. For years, she had watched her neighbors compete for the greenest lawn on the block, never once considering that their vanity was strangling the heart of their home. "They don't even know," she whispered. "They think they're just watering their grass."

"They don't want to know," Micah replied, closing his kit with a sharp, definitive click.

They walked in silence back toward the center of the wooden deck. Declan was hunched over a tablet, the blue light of the screen clashing with the golden hour sun. Zara was pacing the length of the boards, a notebook clutched in her hand like a shield. As Avery and Micah approached, the energy on the deck shifted. There was no more casual banter, no more excitement of a discovery. The air felt heavy with what they now knew.

Declan didn't look up as they arrived. "I finished mapping the hydrostatic pressure in the northern cave system," he said, his voice tight. "When you cross-reference the flow rate drop with the construction timeline of the Enclave's new stormwater retention basin, the math is undeniable. They didn't just cap the surface recharge points. They effectively severed the main artery that feeds the basin."

Zara stopped pacing, looking at the group. "I checked the land-use filings from the last two decades," she said, her voice steady but lacking its usual journalistic bravado. "The permits were rubber-stamped. No environmental impact study accounted for the cumulative stress of the neighborhood expansion. We've been treating the spring like an infinite resource, while simultaneously paving over every sponge the land had to offer."

Avery stood between them, looking down at the dark, shrinking pool. The spring, once a vibrant, breathing entity, looked sluggish, its currents muted and tired.

"We've been looking for a villain," Avery said, her voice gaining strength as she looked at her friends. "We were looking for one big event, one bad actor who broke the spring. But it's not just one person. It's us. It's the cumulative cost of every 'small' decision. The neighbor who fertilizes his lawn, the developer who builds one extra row of houses, the planning commission that doesn't check the maps."

Declan finally looked up, his eyes reflecting the deep concern that had been simmering in him for weeks. "It's a systemic tipping point, Avery. If we don't force a change in the infrastructure—if we don't force them to deconstruct those illegal caps and restore the floodplains—the spring isn't just going to struggle. It's going to stop."

"It's already stopping," Micah added, pointing to a patch of exposed lily roots that were beginning to wither in the sun. "The ecosystem is reacting to the stress. If the chemistry doesn't balance, the vegetation dies. If the vegetation dies, the micro-habitats vanish. We are watching the collapse in real-time."

Zara stepped forward, closing her notebook. She looked at each of them in turn, her eyes hardening with resolve. "The Town Hall meeting is in two days. If we present this as a list of grievances, they'll laugh us out of the room. They'll call it 'teenager alarmism' again, just like they did before. They'll dismiss the data as amateur science."

"Then we don't just present the problem," Declan said, tapping his tablet. "We present the solution. We have the maps, we have the chemical signatures, and we have the historical record. If we link the decline of their property values to the loss of the spring, it's not an environmental issue anymore—it's a community crisis."

Avery looked at the spring, her heart aching for the life beneath the surface. She had spent her life observing the creatures here, learning the language of the water. She realized now that she wasn't just an observer anymore. She was a witness, and the duty of a witness was to ensure the truth was heard.

"It feels irreversible," Avery said, her voice dropping to a whisper. "Sometimes, when I look at how much we've damaged, I feel like we're trying to stop a tidal wave with a screen door."

Micah reached out and placed a hand on her shoulder. His touch was grounding, a calm center in the chaos of their realization. "The spring is resilient, Avery. It's survived droughts and shifts in the earth for thousands of years. It doesn't need us to save it, not really. It needs us to stop strangling it. It needs us to step back and let the water flow the way it was meant to."

"They won't want to hear that they have to change their habits," Zara noted, though her tone was no longer questioning. It was calculating. "They're going to be angry when they realize we've been tracking their water usage and their lawn chemicals."

"Let them be angry," Declan said, his gaze shifting toward the horizon. "Anger is a form of energy. We can use it. We need to make sure we're the ones directing it."

Avery stood a little straighter. The humidity seemed less oppressive now, transformed into the very air she needed to breathe. She looked at her friends—a geologist, a journalist, and a naturalist. They were four kids who had walked into this spring with nothing but curiosity, and they were walking out with a mission that could reshape the future of their entire valley.

"We have two days," Avery said, her voice clear and resonant. "We document everything: every pipe, every permit, every reading. We make the case so clear that they can't look away. We show them the map, we show them the chemistry, and we show them that if we don't act, the Glasswater Spring is going to be nothing more than a memory in a history book."

"It's not just about the spring anymore, is it?" Zara asked, a small, sad smile touching her lips. "It's about whether or not this community is willing to grow up and take responsibility for its own footprint."

Micah looked at the water one last time, his expression one of quiet, fierce hope. "It's about whether or not we are capable of listening to the land beneath our feet. Because right now, the land is screaming."

Avery nodded. She picked up her camera bag, the weight of the equipment suddenly feeling less like a burden and more like a tool of survival. She knew the road ahead was steep. She knew there would be resistance, denial, and perhaps even open hostility from the very people she grew up with. But as she looked at the water—the precious, dwindling water—she knew there was no other choice. They had crossed the threshold of discovery. Now, they were stepping into the work of restoration.

"Let's go," Avery said, turning away from the basin. "We have a lot of work to do."

Chapter Eighteen

THE CUMULATIVE COST

The air inside the humid, low-ceilinged limestone cavern was heavy, thick with the smell of wet minerals and ancient, trapped breath. Declan adjusted the aperture on his specialized sensor array, the blue light of the readout casting long, jittery shadows against the porous walls.

"If the hydrostatic pressure were functioning as a closed loop," Declan murmured, his voice echoing flatly against the damp stone, "we shouldn't be seeing this much atmospheric seepage. Look at the calcification here, Avery. It's brittle. It's parched."

Avery knelt beside him, her headlamp cutting through the gloom. She traced a finger along the jagged, dry edge of a limestone shelf that should have been submerged beneath at least two feet of crystal-clear aquifer water. Instead, it was dusted with a fine, greyish silt—the calling card of stagnant, blocked systems.

"It's not just the drought, Declan," Avery said, her voice steady despite the sinking feeling in her gut. She pulled out her digital notepad, flicking through the time-lapse photographs she'd taken of the vegetation at the spring's mouth over the last week. "Look at the *Hydrilla* and the native bladderwort. They're turning brown at the tips. They aren't just thirsty; they're choking on something. The chemistry in the runoff from the

Enclave isn't just taking the water—it's changing the biology of the conduit."

Declan leaned over, his brow furrowed as he synced his tablet with her device. The data didn't just overlap; it formed a jagged, incriminating map. "Okay, overlay the drone's infrared scan of the surface storm drains with your vegetation stress points."

Avery tapped the screen. As the maps merged, the truth became undeniable. The storm drain network, designed to shunt rainwater away from the Enclave's manicured lawns and paved cul-de-sacs, acted like a surgical bypass. It intercepted the natural surface flow that was supposed to percolate through the porous limestone—the very limestone they were standing on—and funneled it directly into a retention pond that leaked into a different, non-recharge basin.

"We're literally looking at a bypass surgery performed on the ecosystem," Declan whispered, his usual scientific detachment slipping. He ran his hand over the limestone. "They paved over the natural catchment, diverted the rainfall into concrete pipes, and effectively starved the spring's heart. These aren't just 'small changes' to a neighborhood. This is a structural redirection of the entire aquifer's local replenishment."

Avery felt the weight of it then. It wasn't a singular villain, a smoking gun in the traditional sense. It was the cumulative inertia of thirty years: a permit here, a curb-cut there, a decision to prioritize dry lawns over deep water. "Every time they added a new phase to the Enclave, they just extended the drain. They didn't think about the spring because they couldn't see it from their driveways. They treated the groundwater as an infinite resource that just... appeared."

"And now it's gone," Declan finished. He stood up, his knees cracking in the damp silence. He began to pack his sensors with rapid, efficient movements. "We have the data. The hydrostatic drop matches the exact timeline of the last three phases of the Enclave development. The nitrate spikes Micah found? That's the direct result of the runoff being diverted from natural soil filtration into these high-velocity storm pipes. It's an open-and-shut case of hydrological negligence."

Avery stayed on her knees for a moment longer, looking at the dry, calcified walls. She imagined the water—the living pulse of the spring—that should have been surging through this cave, carrying the light of the surface down into the deep. "We have the proof, but people don't like to hear that their way of life is the problem, Declan. Especially when it's something as mundane as how they maintain their landscaping."

"That's why we don't present it as a blame game," Declan said, his tone shifting from clinical to determined. He held out a hand to help her up. "We present it as a system failure. It's an engineering error. If we can show them that restoring the natural drainage would actually stabilize their own property values and prevent the flooding the Enclave has been having, we turn their self-interest into our leverage."

Avery stood, brushing the limestone dust from her pants. The quiet teen who usually preferred the company of her camera lens over a crowd suddenly felt the gears shifting in her mind. She wasn't just observing the decay anymore; she was documenting the path to recovery.

"We need more than just the charts," she said, her voice growing firmer as they walked toward the narrow, light-filled exit of the cave. "We need to show them what they're losing. The photography I have of the birds leaving, the changes in the water color... it needs to be part of the synthesis. People don't react to flow rates, Declan. They react to the story of the spring going silent."

"Then let's write the story," Declan replied, ducking his head as they emerged from the cave into the humid, blinding Florida afternoon.

The heat hit them like a physical weight, but for the first time, it didn't feel oppressive. It felt like the environment they were fighting to save. They stood on the edge of the basin, the spring a ghost of its former self, its surface glassy and sluggish.

Avery held her camera, but she didn't take a picture. Not yet. She stared at the water, seeing not just the lack of flow, but the potential for it to return.

"We have to be clinical," she reminded herself, though her pulse was racing. "We have to show the data, explain the conduit, and then show the impact. Micah and Zara are going to have to do the heavy lifting with the community. We need to give them the ammunition."

"The ammunition is ready," Declan said, tapping his tablet. "I've mapped every blockage. Every point where the concrete meets the limestone. We show the town how to fix the plumbing of their own backyard."

They walked back toward the park entrance, their footsteps heavy with the burden of the truth they now carried. They were moving from the role of curious wanderers into architects of a restoration, the weight of the last three decades of human error resting on their shoulders, and for the first time, they felt strong enough to carry it. They weren't just teenagers in a park anymore; they were the first responders to an environmental slow-motion catastrophe, and the spring was waiting to see if they were enough to save it.

The research shed smelled of damp earth and stale coffee, a frantic sanctuary pinned together by pushpins and red string. Micah sat hunched over a spread of satellite imagery, his finger tracing the jagged blue line of the creek that once fed the marshland—an artery now choked by concrete curbs and storm drains.

Zara paced the narrow aisle between the drafting table and the window, her phone pressed to her ear, listening to the hollow drone of a county official's voicemail. She hung up, a sharp, metallic *click* that echoed against the corrugated metal walls.

"They're stalling, Micah," she said, her voice tight. "The planning commission didn't just 'forget' to file the impact report for the Enclave's phase two. It's sitting on a desk somewhere, gathering dust because they don't want us to see the original grading permit."

Micah didn't look up. He was overlaying a map of the current nitrate spikes—the ugly red blotches on his printouts—onto a historical survey of the old citrus groves. "It's not just a missing file, Zara. It's a pattern. Look at the elevation data from thirty years ago. The natural flow wasn't just 'blocked' by the construction of the Enclave; it was re-engineered. They essentially turned the entire drainage basin into a funnel aimed directly at the municipal sewer, bypassing the natural karst windows. They didn't just build on the land; they lobotomized the aquifer's ability to recharge."

Zara stopped pacing, leaning over the table to inspect the maps. The frustration that had fueled them for weeks flickered, but beneath it, the weight of the realization began to settle. They weren't just fighting a developer's hubris or a bit of bad zoning. They were unraveling forty years of convenience.

"This isn't about one bad actor," Zara whispered, her journalist's brain shifting gears. "It's about the whole town. Every driveway, every manicured lawn, every HOA rule that demands a perfect green patch of grass in the middle of a sandy ridge. We've been starving the spring to keep our yards pretty."

"And the community doesn't know," Micah said, his tone heavy with a mix of sadness and resolve. "They think the spring is a faucet that can't run dry. They don't see the connection between their fertilizer and the algae bloom at the vent, or their driveway runoff and the silence in the Grotto."

Zara grabbed a legal pad, her pen hovering over the paper. "If we just throw charts at them, they'll tune us out. People don't vote for spreadsheets, Micah. They vote for the things they love. We have to tell the story of the water. Not as an abstract chemical formula, but as a resident. A neighbor."

"A neighbor that's currently dying of thirst," Micah added. He looked up, his eyes searching hers. "If we go public with this, we're going to be targets. The Enclave, the local council, even the neighbors who don't want to admit their landscaping is part of the problem. It's going to get messy."

Zara gave a sharp, humorless laugh. "Messy is where I do my best work. But I can't do it alone. I need you to translate the science. I need people to understand that when the spring dies, it's not just about the loss of a swimming hole. It's the loss of the soul of the place. It's the loss of our stability."

Micah reached out, smoothing a crease in the map of the basin. "I can show them the roots. I can show them how non-native ornamentals are choking out the native plants that used to filter the water. If we combine your narrative with my ecological data, it's not an accusation anymore. It's an invitation to fix it."

"An invitation to be better," Zara corrected. She started sketching out a structure for their presentation. "We need to frame this as a choice—a legacy. We tell them about the 'before'—the days when the water was crystal clear, and the flow was a constant, steady heartbeat. Then we hit them with the 'after'—the cumulative weight of our choices. And then, we give them the solution."

"The Recharge Initiative," Micah said, the name sounding more real now, less like a pipe dream. "Restoring the filter zones. Permeable pavement requirements. Mandatory buffer zones around the sinkholes."

"Exactly," Zara said, her pen moving with renewed vigor. "But it has to be framed as a community rescue mission. We aren't pointing fingers at the developers; we're pointing at the problem and asking the community to step up. We turn the outrage into ownership."

The silence in the room changed. The frantic, nervous energy that had plagued them for days shifted into something grounded—a grim, clear-eyed focus. They were no longer just kids digging through trash bins and old records; they were architects of a new direction for the town.

"How are we going to get them to listen?" Micah asked. "The last town hall was a circus of denial."

"We make them see it," Zara said, a glint of determination in her eyes. "We don't just tell them the spring is dying. We show them the pulse. We use Avery's photos. We use Declan's time-lapse of the vent. We put the reality of the Grotto in their faces until they can't look away. We make it impossible for them to pretend that everything is fine."

Micah stood up, his joints popping, his face set in a look of quiet, steady resolve. "Then let's finish the synthesis. I want the nitrate data for the last five years mapped against the expansion timeline. If we can show that the spikes directly correlate to the development phases, we have the smoking gun."

Zara nodded, tearing the page from her legal pad and pinning it to the wall beside a blurry photo of a dry limestone shelf. "And I'll start drafting the article. I'm going to make the community see that this isn't the end of their spring. It's the beginning of a different kind of life—a life where we stop taking from the land and start giving back."

Outside, the humid Florida afternoon began to press against the windows, the heavy air a reminder of the water they were fighting to save. Inside, the two of them worked in silence, the wall of maps and notes slowly knitting together into a narrative that was becoming more than the sum of its parts. It was no longer just an investigation; it was an ultimatum. They were prepared to stand in front of the town and demand that their home be given a chance to breathe, to flow, and to heal.

As the sun dipped low, casting long, bruised shadows across the shed, the scale of what they were attempting finally struck them. They were asking a community to change its habits, its economy, and its very foundation. It was an impossible task. And yet, looking at the maps, the data, and the evidence of the life struggling to survive beneath the surface, they knew they didn't have a choice. The spring was waiting, and it was running out of time.

Chapter Nineteen

FRAGMENTS OF THE TRUTH

The overhead fluorescent light hummed, a jagged, rhythmic buzz that seemed to vibrate in time with the tension in the room. Declan's garage smelled of motor oil, damp limestone dust, and the sharp, ozone tang of cooling electronics.

Avery stood at the center of the workbench, her hands hovering over a sprawling mosaic of evidence. She wasn't looking at the papers; she was looking at the way they intersected. She pushed a faded, sepia-toned aerial map from 1974 next to a high-resolution, false-color thermal scan Declan had pulled from his drone just yesterday.

"Look at the drainage basin," she whispered, her voice barely rising above the garage's drone. "The Enclave development didn't just build houses. They paved over the pulse."

Declan leaned in, his knuckles white as he gripped the edge of the workbench. He tapped a glowing coordinate on his tablet. "It's not just the paving, Avery. It's the depth. When they leveled the north ridge for the retention ponds, they cut into the confining layer—the clay seal that keeps the aquifer pressurized. They didn't just divert the water; they punctured the balloon."

Micah was sitting on a stack of milk crates in the corner, nursing a lukewarm coffee. He had been silent for an hour, his eyes moving back and forth between the nitrate data spreadsheets and a collection of water samples he'd taken from the Grotto. "It's not just one point of failure," he said, his voice quiet, heavy with the weight of the realization. "It's the cumulative effect. The lawns in The Enclave, the run-off from the citrus groves two miles up-gradient, the storm drains... the whole system has been screaming for decades. We were just the first ones actually to listen."

Zara paced the narrow aisle between the workbench and the tool rack, her phone clutched in her hand. She had been drafting leads, deleting them, and starting over. The sheer gravity of what they had assembled—the evidence of a multi-generational, systemic collapse—was paralyzing.

"If we just drop this in the paper," Zara muttered, spinning to face them, "we're not just criticizing a developer. We're telling three generations of homeowners that their 'dream homes' are the reason the spring is dying. We're telling the town council they've been signing off on a slow-motion suicide for the watershed. This isn't a story about a drought anymore. It's an obituary for the status quo."

"Which is exactly why we can't just be 'teenagers with a hobby' anymore," Declan said, his pragmatic edge sharpening. "If we present this as a list of grievances, they'll bury us. They'll call it alarmism, cite 'unforeseen environmental variables,' and wait for the news cycle to move on."

Avery picked up a red marker and drew a circle around the intersection of the primary limestone conduit and the drainage pipe that led to the city sewer. "It's not about blame," she said, finally meeting their eyes. "It's about anatomy. You don't blame the body for the injury; you treat the wound."

She looked at the maps, the chemical logs, and the drone footage—the cold, hard evidence of their small town's unraveling. The air in the garage felt suddenly thin. The shock of the discovery was giving way to a sickening, profound sense of responsibility. They were no longer just observers; they were holding the diagnosis of a patient everyone else insisted was just 'tired.'

"My grandfather told me the spring used to flow so hard you couldn't stand against the current at the mouth," Micah said, breaking the silence. "He never understood the 'why.' He just knew it changed. If we show them the 'why'—if we show them that this is a plumbing problem that can be re-routed, that we can restore the natural filtration, that the sinkholes can be unplugged—"

"Then it's not an obituary," Zara finished, her eyes narrowing as she began to see the structure of the narrative. "It's a recovery plan. We stop being the kids who found a problem, and we start being the consultants who brought the solution."

Declan looked at his maps, then back to the group. The nervous energy that had driven him since the Spring Discovery Day had transformed into something colder, more focused. "It's a massive engineering undertaking," he admitted. "We're talking about tearing up pavement, installing bio-swales, recalibrating the storm-drainage—"

"But it's a tangible, physical goal," Avery interrupted. "We're not asking them to stop living. We're asking them to change *how* they live. We present the connection. We show them the map, we show them the nitrates, we show them the flow velocity charts. We make it impossible to look at their own driveways and not see the spring."

Zara sat down on the edge of the workbench, her posture shifting from anxious to authoritative. "We need to weave the stories. The journals I found—the ones from the original citrus settlers—they talked about the water being 'sweet enough to drink without a filter.' That's our opening. We anchor them in what was lost, show them the data on how it was dismantled, and then lay out the map for what we bring back."

Micah stood up, his coffee forgotten. He walked over to the workbench and placed a hand on the aerial map, right over the dry Grotto. "It has to be inclusive. If we make them the enemy, we lose the town. If we make them the rescuers, we win the spring."

The silence returned, but this time, it was different. It wasn't the silence of confusion; it was the quiet concentration of a team that had just accepted that the map they were holding was the only thing standing between the Glasswater Spring and total, irreversible silence.

Avery picked up the red marker again. She didn't draw a line this time. She drew a circle around the whole region, encompassing the houses, the roads, the sinkholes, and the spring itself.

"We start with the facts," Avery said, her voice steady, stripped of its usual shyness. "No speculation. No accusations. Just the geometry of the ground beneath us. If we show them the science, the truth becomes an invitation, not an indictment."

Declan nodded, his hand hovering over his tablet, ready to refine the model for the town hall presentation. "I can simplify the conduit mapping. Make it look like a flow-chart. Anyone can understand a clogged pipe."

"And I'll write the narrative," Zara said, her fingers already dancing across her phone screen. "One story for the paper, one script for the meeting. We bridge the gap between the 'how' and the 'what now.'"

Micah smiled, a faint, hopeful expression that softened the tired lines around his eyes. "And I'll handle the ecological evidence. If we need to explain why a specific plant or insect is the canary in this coal mine, I'll make sure they understand that it's not just for the sake of the ecosystem—it's for the stability of their own backyards."

Avery looked around the small circle of friends. The garage, with its clutter and its buzzing lights, no longer felt like a hiding place. It felt like a war room. They were four teenagers, standing in the dark, preparing to light a fire that would either illuminate the path to the spring's future or burn their reputations to the ground.

"Okay," Avery said, exhaling a long, slow breath. "Then let's start from the beginning. Not the drought. Not the discovery. Let's start with the water."

As she turned the page of her notebook, the frantic energy of the last few days settled into a deliberate, rhythmic focus. They had the evidence; now, they had to craft the truth.

The air in the garage was thick with the scent of damp concrete and ozone, punctuated by the hum of Declan's high-powered laptop fan. A whiteboard leaned against a stack of tires, covered in a chaotic, spiderwebbing diagram of drainage pipes, historical parcel lines, and the geological spine of the aquifer.

Zara tapped her pen against a notebook, her face illuminated by the harsh blue light of a dozen open browser tabs. "If we go in there and start naming names—calling out the Enclave HOA for the fertilizer runoff, pointing to the '74 zoning records as proof of criminal negligence—the room is going to burn down before we get to the slide on groundwater recharge," she said, her voice tight. "We'll be the teenagers who cried wolf, and they'll shut us down faster than a spring-fed creek in August."

"But the data is the data," Declan argued, gesturing at the printouts of hydrostatic pressure readings spread across a workbench. "If I don't show them the physical conduit blockages—the fact that their 'landscaping improvements' are literally choking the aquifer—they'll treat it as a matter of opinion. We have to show the cause. The cause is them."

Micah, who had been sitting on the floor cross-referencing a map of native cypress root systems with a modern development blueprint, looked up. He didn't look angry, just profoundly tired. "Declan, if you walk into that Town Hall and tell the people who live in the Enclave that they've effectively murdered the spring with their lawn sprinklers, they're going to stop listening the second the word 'murder' hits the air. They'll get defensive. They'll hire lawyers. They'll build a wall of denial that we can't climb over."

Avery sat on an overturned crate near the edge of the garage, watching the four of them reflected in the dark, oily surface of a drainage pan they'd used for an experiment. She traced the path of a stray bead of condensation running down the side of a water sample bottle.

"Micah's right," Avery said, her voice quiet but carrying clearly through the narrow space. "When I'm out on the water, if I see a bird acting erratic, I don't yell at it to stop. I look at why it's reacting that way. The people in this town... they think they're doing the right thing. They think a green lawn is pride. They think a paved drain is progress. They aren't trying to destroy the spring. They're just disconnected from it."

"So, what? We coddle them?" Zara asked, though her tone had softened. "We have the proof that their 'pride' is drying up our heritage. Do we bury that?"

"No," Avery stood up, moving to the whiteboard. She picked up a dry-erase marker and circled the entire sprawling web of evidence they had spent weeks gathering. "We don't hide the truth. We change the story. Instead of an autopsy, we offer them a homecoming."

Declan frowned, leaning back in his chair. "A homecoming?"

"The spring isn't just dying," Avery continued, her eyes bright as she looked at her friends. "It's waiting. Everything Micah found—the nitrate filtration, the native ground cover—that's the roadmap for how the spring *wants* to be. We show them the maps of how it was before the developments. We show them the photos of the clarity, the wildlife, the way the flow used to cycle. We make them feel the loss, not as a crime they committed, but as a treasure they've forgotten how to keep."

Zara stopped typing. She looked at the screen, then at the wall of evidence. "The 'Glasswater Recharge Initiative.' It sounds like a recovery program, not a court summons."

"Exactly," Micah nodded, standing up to join Avery at the board. "We frame it as a collective project. We show them that their property values depend on the health of the aquifer. We show them that their 'lawn pride' is actually a tax on their own future water security. We give them the 'how-to' guide for fixing it. Not 'stop doing this,' but 'here is how you can help bring the water back.'"

Declan looked at his maps, then at the complex chain of causation he had mapped out—a chain that clearly led back to the doors of the wealthy residents who would be sitting in the front rows. He let out a long, slow breath. "It's a massive pivot. I have to strip out the 'blame' slides. I have to focus on the hydro-engineering solutions—how simple swales and native plant buffers can re-route that surface runoff back into the sinkholes. It's a lot of work to reformat."

"It's the only way we get a vote," Zara said, a sharp, determined smile breaking across her face. "And I can work the narrative. 'Restoring the Heart of the Spring.' I'll frame it as a community-wide legacy project. We aren't here to expose a villain; we're here to lead a renewal."

"And I'll handle the ecological evidence," Micah added, picking up a stack of index cards. "I'll focus on the benefit to them. Pollinator gardens, flood mitigation, water conservation—things that actually make their lives better. We show them that a healthy ecosystem is a luxury they haven't realized they're paying for."

The tension in the garage seemed to evaporate, replaced by a frantic, focused energy. They were no longer just observers or investigators; they were architects of a new reality. They worked through the night, the sound of crickets outside the garage door a rhythmic reminder of the wild world they were fighting to save.

Avery stayed by the whiteboard, refining the visual flow of their presentation. She thought back to the very first day, the moment the water in the spring had flickered and dimmed, and how helpless she had felt. She looked at her hands—covered in dry-erase marker dust—and then at the team.

"We're going to give them a choice," Avery whispered. "We're going to make the right path the most attractive one."

"It's a gamble," Declan said, pausing his typing to look at her. "If they reject the solution, we're back to square one with a very angry neighborhood."

"They won't," Zara said firmly, not looking up from her laptop. "Not when we present it as a rescue mission. People love being the heroes of their own story, Declan. We're just giving them the script."

They settled into a rhythm, the clatter of keyboards and the squeak of markers against the board defining the next several hours. They were no longer documenting the decline; they were drafting the blueprint for the recovery. As the first gray light of dawn began to bleed through the dusty garage windows, the whiteboard was no longer a tangle of crimes and consequences. It was a clean, logical path forward—a roadmap that connected the deepest, hidden secrets of the aquifer to the front porches of every house in town.

Avery walked to the garage door and rolled it up, letting in the cool, humid air of a Florida morning. Outside, the world was waking up. In the distance, she could almost imagine the rhythmic, steady pulse of the spring, waiting for the help they were about to provide.

"We're ready," she said, turning back to the room.

The others stopped. They looked at each other—a geologist, a journalist, a naturalist, and an observer—and in that moment, the weight of the last few weeks settled into something manageable, something they could carry together. They had the evidence, they had the plan, and for the first time, they had the conviction that the spring didn't just have a past—it had a future.

Chapter Twenty

A Collaborative Conclusion

The air inside the converted storage shed behind Micah's barn was thick with the scent of damp soil and the sharp, ozone tang of cooling electronics. Every flat surface was occupied: topographic maps weighted down by heavy river stones, stacks of yellowed zoning records, and Declan's laptop, its screen casting a flickering, sterile blue light across their faces.

They were beyond exhaustion. The adrenaline of the confrontation at the Enclave had burned off hours ago, replaced by a dull, throbbing weight in their temples.

Avery stood at the edge of the workbench, her fingers tracing the contour lines on a topographic printout. She could feel the spring—not the water itself, but the *absence* of it, like a phantom limb. She had spent the last three hours trying to reconcile the historical flow charts with the current drainage reality.

"It's not just the paving," Avery said, her voice raspy. She didn't look up. "If we go into that Town Hall and just talk about the parking lot, they'll call it an aesthetic grievance. They'll offer to plant a few trees and call it a day."

"She's right," Declan muttered, leaning back, his chair creaking in the small room. He looked at his drone's flight path logs, a tangled web of red lines on his screen. "The hydrological connection is the anchor. If we don't explain how the conduit under the North Bluff acts as a literal siphon for the aquifer, the argument doesn't hold water. Pun intended."

Zara paced the narrow aisle between the workbench and the shelving unit. She held a stack of index cards, her thumb flicking against the edges rhythmically. She was the one who had to weave this into something the town council could understand—something that couldn't be dismissed as teenage alarmism.

"The narrative isn't 'The spring is dying,'" Zara said, stopping to stare at them. "The narrative is 'The town is losing its foundation.' We have to frame the aquifer as a shared utility. If we lose the spring, we lose the water pressure in their homes. We lose the property value of the homes near the park. We make it personal, and we make it expensive."

Micah sat on a stool, his hands wrapped around a lukewarm mug of herbal tea. He had been quiet, letting the others hash out the tactical data. Now, he looked up, his expression one of measured, steady calm.

"You're both approaching this like engineers and journalists," Micah said softly. "But the people in that room tomorrow? They're parents. They're business owners. They're scared of change. If we come at them with a list of demands, they'll put their guard up. We need to invite them to be the heroes of this story."

Avery turned, eyes narrowing slightly. "Inviting them to be heroes doesn't change the fact that they've been ignoring the runoff for a decade, Micah."

"No," Micah agreed, "but it acknowledges that they didn't do it on purpose. It wasn't malice, Avery. It was ignorance. We were ignorant too, until we started looking. We have to meet them where they are."

Declan shoved his laptop aside, the sudden movement sending a stack of printouts sliding toward the floor. Avery caught them just in time.

"I don't have the luxury of being diplomatic, Micah," Declan snapped, his frustration finally fracturing his composure. "I have the core samples. I have the nitrate levels from the runoff. I have proof that the Enclave is dumping chemical-heavy sludge directly into the limestone fissures that feed the Glasswater headspring. If I present this as a 'mistake,' I'm lying."

"You aren't lying," Zara interjected, stepping between them. "You're framing. There's a difference." She turned to the whiteboard, grabbing a dry-erase marker. "Declan, you lead with the science—the 'what.' Avery, you provide the 'why'—how the biology of the spring has fundamentally shifted. I'll handle the history—the 'how we got here.' And Micah? You're the bridge. You show them that the recovery isn't a pipe dream. You show them the future we can have."

The room fell silent. The ticking of a wall clock seemed unnaturally loud.

Avery walked to the whiteboard and drew a simple, jagged line representing the water table. She watched the black ink bleed into the white surface. She thought about the way the water had looked three days ago—cloudy, sluggish, choked with algae that shouldn't be there. She remembered the heron that had left the basin, realizing the food source was collapsing.

"It's not just about saving the water," Avery said, her voice dropping to a low, intense register. "When I was out there this morning, before the rain, I sat by the grotto. I realized that if we don't fix this, the land around the spring will die too—the oaks, the cypress, the gopher tortoises. We aren't just protecting a tourist attraction. We're protecting a living lung."

Declan looked at her, his posture softening. The rigidity in his shoulders seemed to evaporate. He looked down at the map of the limestone conduit, then back at his friends.

"Okay," Declan said, his voice quiet. "Okay. We focus on the connectivity. The aquifer doesn't care about property lines. The water doesn't care who signed the zoning permit in 1974. We present it as a closed system. If one part suffers, the whole system fails."

Zara started writing on the board, her handwriting sharp and decisive.

1. The Warning: The Current Status.2. The Culprit: Infrastructure, not Weather.3. The Consequence: Property and Sustainability.4. The Solution: The Reconstruction Initiative.

"It's a tightrope," Zara murmured. "We have to be clinical enough to be credible, but passionate enough to matter."

"We have the truth," Micah said, standing up and setting his mug aside. "That's all we've ever had. We have to make sure they hear it."

Avery looked at the group. For a moment, the weight of the last few weeks felt lighter. They weren't just four kids with a hobby anymore. They were a team, each holding a different piece of a mirror, and finally, they were aligning them to show the town the reflection of what they had done.

"We need a rehearsal," Avery said, grabbing a stack of index cards from the table. "Zara, you're the moderator. Declan, you're the expert witness. I'm the field researcher. Micah... you're the community voice. We do this until we can say it in our sleep."

Declan pulled his chair closer to the table, his fingers already dancing across his keyboard to pull up the slides for their presentation. "Start from the beginning. No jargon. If I use a geological term, someone better hit me."

"Understood," Zara said, a small, confident smile touching her lips. "Let's start from the moment the flow stopped."

Outside, the first real rain of the season began to drum against the tin roof of the shed. It was a steady, rhythmic sound—the sound of the aquifer beginning to remember how to breathe. They worked through the night, the blue light of the screens the only sun they knew, turning their fear and their exhaustion into a weapon of logic and hope.

As the first gray light of dawn began to bleed through the cracks in the walls, the presentation was finished. It was lean, it was brutal, and it was undeniably true.

"We're ready," Avery whispered, leaning back. Her eyes burned, but for the first time in weeks, she felt entirely, terrifyingly awake.

"We're ready," the others echoed, the weight of the task ahead finally balanced by the strength of the three people sitting beside them. They were ready to face the town, and they were ready to save the spring.

The high school gymnasium smelled of floor wax and humid, trapped anticipation. The fluorescent lights hummed, an uneven strobe over the faces of nearly two hundred residents. Avery stood in the wings, her fingers tracing the rough edge of a geological core sample tucked into her pocket. The air felt heavy—a physical manifestation of the tension in the room.

Zara stood at the podium, adjusting the microphone. She looked every bit the seasoned investigator, even if her dress shoes were pinching her feet. Behind her, the projector displayed a map of the county, overlaid with the jagged, damning red lines of the Enclave's unauthorized drainage bypasses.

"We aren't here because of the weather," Zara began, her voice steady, echoing off the rafters. "We are here because of choices. Choices made in boardrooms, offices, and backyards over the last fifty years."

A rumble of discontent moved through the crowd. Mr. Henderson, a local developer who had been the loudest critic of their "teenage alarmism," leaned forward in the front row, arms crossed tightly.

Zara didn't flinch. She gestured to the screen, where the 1974 zoning covenant flickered to life. "This document, which has been gathering dust in the county archives, wasn't just a suggestion. It was a binding agreement. When the Enclave was built, they were granted permits on the condition that they maintained the natural sheet flow of the northern basin. They didn't. Instead, they built walls. They buried the limestone conduits."

"That's ancient history," someone shouted from the back. "What about the rain? We've been in a dry cycle for two years!"

Micah stepped forward, moving with a calm, practiced ease that seemed to drain the heat from the outburst. He didn't look at the dissenter; he looked at the crowd as a whole, his hands open.

"The rain isn't the problem," Micah said softly. "The soil is. When I walked the Enclave perimeter this morning, I didn't see native palmettos. I saw perfectly manicured turf grass—a sponge for nitrogen-heavy fertilizers. When it does rain, that water doesn't soak into the aquifer to recharge the spring. It washes the toxins into the drainage pipes, bypassing the limestone filters entirely. We're starving the spring and poisoning it at the same time."

He held up a vial of water taken from the spring's main vent. It was cloudy, a murky green compared to the clear, crystalline water the town remembered from their youth. "This is what we're drinking. This is what the manatees are trying to survive in. It isn't a drought. It's a blockage."

Declan walked to the projector, plugging in his laptop. He didn't try to charm the audience; he showed them the ground beneath their feet. He clicked a button, and a 3D rendering of the subterranean cave system appeared. It was complex, beautiful, and utterly scarred.

"I spent forty-eight hours mapping these conduits," Declan said, his voice clipped and efficient. He pointed to a series of darkened bottlenecks in the model. "This is a limestone fracture. It's the heart of the spring's flow. When the Enclave diverted the surface water, they didn't just move a river; they cut off the circulation to the entire aquifer. If we don't remove these blockages, if we don't restore the natural topography, the spring won't just stop flowing—it will collapse. The sinkholes won't be a curiosity anymore. They'll be a liability for every property owner within five miles."

The room went deathly silent. The mention of property values did what the environmental plea couldn't: it made the threat personal.

Avery stepped up to the microphone. She felt small, but as she looked out at the faces of the people she'd seen at the spring every weekend of her life—the kayakers, the families, the elderly couple who sat on the bench under the live oak—she felt her voice solidify.

"I've spent every weekend since I was seven at Glasswater," Avery said, her voice barely above a whisper, yet carried clearly by the mic. She didn't look at her notes. She looked at the center of the room, at the people who had dismissed them. "I've watched the way the light catches the water at 7:00 AM. I've seen the way the fish hover near the boil when the water is healthy. For the last six months, I've watched that life fade. It's not just a resource. It's the reason this town exists."

She paused, taking a breath. "The scientists tell us the mechanics. The journalist tells us the history. But I'm telling you the consequence. We are the ones who chose the lawn fertilizer over the spring. We are the ones who allowed the developers to build over the conduits. We can either be the generation that let the heart of this town go dry, or we can be the ones who finally decided to listen to the land."

She walked back to the group. The silence stretched, thick and suffocating.

Mr. Henderson stood up. The room held its breath. He looked at the map, then at the vial of water in Micah's hand, and finally at the four of them. He wasn't smiling, but the aggression had drained out of his posture, replaced by a weary, begrudging recognition.

"You're asking us to tear up the North Basin?" Henderson asked.

"We're asking you to do what was promised in 1974," Zara replied instantly.

"It'll cost a fortune," Henderson muttered, but he wasn't looking at them anymore. He was looking at his neighbors, some of whom were nodding slowly.

"The cost of a dead spring," Micah added, "is much higher than the cost of a few drainage pipes."

The town council chair, a woman who had been notoriously neutral throughout the entire ordeal, leaned into her microphone. "We have a motion on the floor to ratify the Glasswater Recharge Initiative. It includes the immediate removal of the North Basin blocks and a permanent moratorium on high-nitrate landscaping in the runoff zone. All in favor?"

One by one, the council members raised their hands.

Avery felt a sudden, dizzying sense of relief as the vote became unanimous. The applause was tepid at first, then swelled into something real—a collective acknowledgment that they had been pushed into a corner by their own neglect and, for the first time, had chosen the harder, right path.

As the meeting dissolved into a flurry of frantic, whispered conversations and angry phone calls from developers, the four of them drifted to the back of the gym.

"They actually did it," Declan whispered, staring at his laptop screen. "They actually passed it."

"It's just a vote," Zara said, though she was smiling, her pen clutched tightly in her hand. "The real work starts when the excavators show up. We have to make sure they follow the blueprints."

"They will," Micah said, glancing at Avery. "They saw the evidence. They can't unsee it now."

Avery looked toward the exit, toward the darkness of the Florida night. She could almost hear it—the distant, rhythmic pulse of the spring, waiting for the pressure to equalize, waiting for the water to find its way home.

"We aren't just researchers anymore," Avery said. "We're the ones holding the map."

"And we're not letting go," Declan added.

Outside, a light rain began to tap against the gymnasium windows. It wasn't a storm, but it was enough. For the first time in months, the sound didn't fill Avery with anxiety about runoff or flooding. It filled her with the quiet, persistent promise of a recharge. The work was beginning, but as she stood there with her team, she knew the spring would be ready for it.

Chapter Twenty-One

Restoring the Source

The silence in the park the morning after the Town Hall meeting wasn't the eerie, stagnant quiet of a dying spring; it was the hum of impending labor.

Avery stood on the edge of the parking lot bordering the Enclave, her boots crunching into the asphalt. Beside her, the heavy mechanical whine of a backhoe echoed through the trees. It was a jarring, discordant sound against the backdrop of the Florida scrub, but to Avery, it sounded like a heartbeat returning to a rhythmic pulse.

"You realize," Declan said, wiping grease and dust from his forehead, "that once we peel back this layer, we're committed. There's no undoing the 'before' of this project."

He gestured toward the section of the parking lot they had spent all week surveying. Beneath this layer of heat-trapping, impermeable concrete lay the calcified remains of a natural swale—a biological kidney the developers had smothered forty years ago.

"That's the point, Dec," Zara said, stepping out from behind the cab of the backhoe. She was holding a tablet, her hair tied back in a messy, practical knot, her clothes dusted with the fine white powder of pulverized limestone. She didn't look like an investigative journalist today; she looked like a project manager. "The town council gave us the green light because we proved the flow is tied to this specific topography. If we stop now, the next heavy rain washes more topsoil and nitrates into the sinkholes. We have to finish the infiltration trench."

Micah approached them, carrying a crate of native sedges and wiregrass, his hands caked in rich, dark compost. He looked exhausted, his usual calm demeanor strained by the sheer physical toll of the last three days. Volunteers from the neighborhood—people who had argued against them only a week ago—were scattered across the lot, some with pickaxes, others with shovels, clearing away decades of debris.

"We've got a problem with the northern drainage gate," Micah said, dropping the crate. "The concrete is thicker than the 1974 blueprints suggested. It's taking the crew twice as long to breach it."

Avery looked toward the obstruction. "The blueprints were probably wrong on purpose. To hide the diversion."

"Whatever the reason," Declan sighed, "it's a physical reality now. I need to get in there with the heavy chisel attachment. Avery, I need you to keep eyes on the flow-rate sensor we dropped in the Grotto. If we shift this mass, I want to make sure we don't accidentally cause a localized collapse."

Avery nodded, already moving toward the boardwalk. The transition from the boardroom to the mud-pit had been swift, but it felt right. For weeks, they had lived in the abstract—in data points, nitrate levels, and historical records. Now, the work was tangible. It was the smell of damp earth and the ache in her shoulders.

She reached the Grotto, where the water remained sluggish, a murky, stagnant reflection of the canopy above. She knelt, checking the digital readout on her sensor. The numbers flickered: *0.4 cubic feet per second.* Still dangerously low, but holding steady.

She heard a splash. A small turtle, an Eastern box, crested the water's edge near a pile of discarded asphalt chunks. It hesitated, then scrambled toward the newly exposed, soft soil where the crew had begun to clear the runoff path. It was a small thing, but it made her breath catch. Life was already testing the edges of their progress.

"Avery!" Zara's voice carried over the sound of the backhoe. She was jogging down the wooden ramp, looking frantic. "We've got a situation. Mr. Henderson showed up with a city inspector. He's claiming the noise and the 'unauthorized excavation' are violating the homeowners' association noise ordinance."

Avery stood up, brushing the dirt from her jeans. "He's trying to stop the work?"

"He's trying to drag us into another hearing," Zara said, her eyes flashing with a familiar, sharpened intensity. "He thinks if he stalls us through the weekend, he can get an injunction by Monday."

Avery looked at the backhoe, then at the sky. The humidity was thickening; the air smelled of ozone. A classic Florida afternoon storm was brewing to the west. "He doesn't understand," Avery said, her voice quiet but firm. "If we don't open that drainage gate before the rain hits, the runoff from his development is going to overflow the storm drains and dump straight into the grotto again, undoing everything we've done."

"I told the inspector that," Zara replied, "but he's looking at the letter of the law, not the geography."

Avery turned toward the work site. Micah and Declan were standing in a standoff with the inspector, a man who looked like he wanted to be anywhere but under the sweltering Florida sun.

Avery walked toward them, feeling a strange surge of confidence. She hadn't just studied the maps; she had felt the pulse of the water beneath her feet for her entire life. She understood the land in a way that couldn't be quantified in a zoning ordinance.

"Excuse me," Avery said, stepping between Micah and the inspector. She didn't raise her voice. She didn't need to. "Sir, I'm Avery Morales. I've been documenting the hydrologic recharge of Glasswater for the last three months. I have the updated drainage topography on my tablet."

The inspector sighed, pulling out a handkerchief to wipe his brow. "Kid, I don't care about the science project. I have a complaint regarding heavy machinery noise before 8:00 AM."

"And if you stop us," Avery said, pointing to the dark, bruised clouds gathering on the horizon, "that noise is going to be replaced by the sound of a sinkhole collapse. Look at the data." She held out her tablet, showing the predictive model Declan had built. It showed the exact route the stormwater would take if the gate remained blocked. "You are legally liable if this construction debris blocks the natural recharge path during the storm. You're looking at an environmental disaster, not a noise violation."

The inspector hesitated. He looked at the clouds, then at the tablet, then at the massive, partially demolished concrete barrier.

"We're trying to fix a forty-year mistake," Micah added, his tone soft but insistent. "The community wants this. Look at them."

He gestured to the surrounding volunteers. A retired teacher was hauling away rebar; a group of parents was laying down native sod to stabilize the bank. They weren't just onlookers anymore; they were stakeholders.

The inspector looked around, clearly sensing the shift in the room—the air was heavy not just with humidity, but with a collective, stubborn purpose. He looked back at Henderson, who was hovering near his sedan, looking increasingly uncomfortable.

"If I sign off on this as an emergency mitigation project," the inspector muttered, "the noise ordinance is suspended until the threat passes."

"Do it," Zara said, holding out a pen.

He took it. As he scribbled his signature, a low rumble of thunder rolled across the spring, shaking the ground.

"Get to work," he said, tucking the clipboard under his arm and turning toward his truck.

Declan didn't wait. He signaled the operator, and the backhoe roared to life, the claw biting deep into the last section of the concrete wall. With a sharp, crystalline crack, the barrier shattered.

For a heartbeat, there was nothing. Then, a wet, sucking sound echoed from the depths of the trench. The water, previously blocked by the concrete, began to move. It wasn't a torrent, but it was a surge—a clear, cold pulse of groundwater finally finding its path back to the surface.

Avery felt it in her toes, a vibration in the bedrock. The spring sighed.

"Look!" Micah shouted, pointing to the Grotto.

The murky water at the bottom of the basin began to churn. A swirling vortex of light-colored sand appeared near the limestone vent. The stagnant film on the surface was pushed aside by the force of the water rising from below.

The rain began to fall—not a drizzle, but a heavy, cleansing downpour. The team stood in the middle of the parking lot, soaked, their hair matted to their foreheads, watching the drainage swale work exactly as they had designed it. The runoff from the road didn't rush into the basin; it diverted into the new, permeable gravel channels, filtering through the layers of sand and native soil they had painstakingly laid.

The groundwater level, visible through the transparency of the rising spring, began to tick upward on the sensor.

"It's working," Declan breathed, his voice barely audible over the rain.

Avery looked at her friends. They were muddy, exhausted, and bruised, standing in the middle of a construction site that had once been a symbol of their community's failure. Now, it was a living monument to their success.

"It's not just working," Avery said, a rare, genuine smile breaking across her face. "It's breathing."

She watched as the first fish—a tiny, silver minnow—darted out from beneath the roots of a cypress tree, darting into the newly clarified center of the spring. It was just a small life in a vast, complex ecosystem, but it was enough. The restoration had begun.

The humidity of a Florida afternoon usually felt like a physical weight, but today, as Avery stood at the edge of the basin, the air felt crisp, almost vibrant. She adjusted her camera lens, the glass cold against her palm. Beside her, the water—once a stagnant, murky amber—had begun to shiver. It was the movement of a heartbeat returning to a dormant chest.

"Look at the eelgrass," she murmured, pointing toward a submerged patch that had looked like dead twine only a week ago.

Micah knelt beside her, his hands caked in the rich, dark muck of the riparian zone. He didn't look up immediately, his focus entirely on the delicate sway of the stems. "It's stabilizing," he said, his voice quiet, stripped of the frantic edge that had defined their late-night planning sessions. "The sediment load from the Enclave run-off has dropped significantly since the bio-retention basins were finished. The spring is finally filtering its own breath again."

A hundred yards away, the rhythmic *thrum* of a hydraulic drill finally cut out, replaced by the shouts of volunteers and the clatter of metal equipment. Declan was standing knee-deep in the secondary drainage channel, his face smeared with grease, holding a tablet up to the sunlight to calibrate the pressure sensor he'd installed near the limestone conduit.

Zara jogged over from the staging area, her notebook tucked into her back pocket. She looked exhausted, her hair pulled back in a chaotic knot, but her eyes held a steady, burning intensity.

"The update from the water board just hit," Zara said, breathless. "They're officially codifying the new land-use regulations. No more aggressive clear-cutting for the next phase of the development. It's done, guys. It's actually in the minutes."

Declan waded out of the water, splashing onto the limestone ledge. He caught the tablet and beamed, a genuine, wide-mouthed grin that looked foreign on his typically stoic face. "The pressure readings are holding. Ever since we cleared that concrete plug near the northern gate, the hydrostatic head has been climbing steadily. We aren't just seeing a spike from the rain—we're seeing a sustained recovery."

"It's not just the water," Avery added, pointing her camera toward a flash of blue near the reeds. A little blue heron landed, stalking the shallows with a precision that hadn't been seen in months. "The hunters are coming back. The ecosystem is recognizing the health of the base layer."

They stood in a small circle, the four of them, looking out over the basin. It had been weeks of back-breaking labor—moving stone, hauling native plantings, and navigating the bureaucratic minefield of the Town Hall—but standing here, it felt like the spring was responding to them personally.

"We have to do a final check of the secondary spillway before the evening squall," Micah said, rising to his feet and wiping his hands on his cargo shorts. "The forecast calls for two inches tonight. If the new basins hold through that, we've proven the system."

"You're acting like a proud father," Zara teased, bumping his shoulder with her own.

"It's the first thing we've built that actually gives back," Micah replied, his tone thoughtful. "Everything else—the roads, the houses—it just takes. This? This is a cycle. We just helped it start turning again."

As they walked toward the northern gate, the sky began to bruise with purple clouds. The Florida storm season was notoriously temperamental, but today, the rain felt like a blessing rather than a threat.

Declan led the way, his stride confident. "The geological data is going to be the most important part of the report I'm writing for the University," he said. "People think the ground is static, that it's just something you build on top of. I think I've got enough evidence now to prove that the karst topography of this region is effectively a living, breathing circulatory system. If you constrict it, the landscape suffocates."

"Make sure you mention the community effort, too," Zara reminded him, clicking a pen. "The data won't mean much if people don't understand that they have the agency to change the outcome. I want this to be a blueprint, not just a case study."

They reached the gate, where the water was already beginning to surge against the newly cleared limestone channels. The sound was a low, melodic roar—the song of a healthy spring. Avery felt it in her chest, the familiar frequency she had missed during the dry spell. She raised her camera and framed a shot: the water cascading over the natural rock, the green of the new vegetation, and in the background, the distant, shimmering heat of the Florida sky.

Avery listened to their banter, a warmth spreading through her chest that had nothing to do with the humid Florida heat. She looked back at the spring one last time. The water was no longer the flat, lifeless mirror it had been a month ago; it was crystalline, revealing the white-sanded floor where the ancient limestone vent pulsed like a steady, rhythmic heart.

They reached the northern spillway just as the first fat, heavy drops of rain began to pelt the canopy. The sound was a symphony—the drumming on the palmetto fronds, the hiss of water against the newly laid limestone riprap, and the gurgle of the drainage system they had painstakingly reconstructed.

"Check the primary intake," Declan shouted over the rising wind, gesturing to the grate they had installed to catch sediment while allowing water to pass.

Avery knelt, shining her waterproof flashlight into the darkness of the trench. The water moving through the channel was clear—a massive improvement over the cloudy, runoff-laden slush that had choked the system for years. She checked the flow gauge attached to the iron grate.

"Flow rate is optimal!" she called out, her voice bright with relief. "The sediment trap is doing its job. No clogging."

"And the nitrate levels?" Micah asked, leaning in close, his face illuminated by her light.

Avery pulled up the latest sensor reading on her phone. "Down by forty percent from last week. The buffer zone is working, Micah. The sedges are doing exactly what you said they would."

The rain intensified, turning the sky a deep, bruised charcoal, but none of them moved. They stood under the shelter of a wide-spreading oak, watching their work hold. The storm, which would have once been a disaster—flooding the roads, washing toxic fertilizers into the spring, and burying the Grotto in silt—was now a source of renewal. It was feeding the aquifer, replenishing the very heart of the spring.

"We did it," Zara said, her voice unusually soft. She reached out and rested a hand on the mossy railing of the boardwalk. "We actually pulled it off."

"We didn't just 'do it,'" Declan corrected, shaking his head as he watched the water swirl into the recharge zone. "We changed the perspective. The town council, the homeowners... they don't see the spring as a property feature anymore. They see it as a lifeline."

Avery felt a profound sense of closure, but more than that, a sense of beginning. She raised her camera and framed the scene: her friends, mud-streaked and exhausted, framed by the lush, rain-drenched green of the spring. She wasn't just observing anymore; she was part of the narrative. She pressed the shutter. The click was tiny, swallowed by the roar of the rain, but the image was captured.

"What now?" Micah asked, looking at the rest of the team. "The restoration phase is a success, but this isn't a 'set it and forget it' situation."

"Monitoring," Zara said, her journalist brain already shifting gears. "We need to maintain the logs. We need to keep the data public. If people stop looking at the spring, they stop caring. We have to keep the conversation going."

"I'll handle the bi-weekly flow reports," Declan agreed. "And I want to expand the drone mapping to the entire northern ridge. If we can map the subsurface cave connections more accurately, we can prevent future developments from encroaching on the main conduits."

Avery looked at the water. "I'll stay with the photography and the wildlife observations. The animals are the best indicators of whether we're actually succeeding or just putting on a good show. As long as the heron stays and the turtles are nesting, we know we're on the right track."

Micah smiled, a slow, genuine expression that reached his eyes. "And I'll keep the native garden project running. I've already got three more neighbors asking how to rip out their sod and plant something that actually helps the water table."

The storm broke as quickly as it had arrived, leaving the park shimmering in the sudden, golden sunlight of a late afternoon Florida sky. Steam rose from the hot asphalt, curling into the cool, damp air of the forest.

Avery walked out onto the wooden dock, the boards still slick with rainwater. She leaned over the railing, looking down into the Grotto. A shadow flickered beneath the surface—a large, dark shape moving against the current. She held her breath.

A manatee, a slow-moving, graceful giant, broke the surface a few yards away, its nose barely brushing the cool, clear water before it submerged again. It was a sign. The spring was open, the connection to the river system was unobstructed, and the habitat was safe once more.

"Hey!" she called out, gesturing to her friends.

They gathered at the railing, their faces lighting up as they saw the visitor. For a long moment, nobody spoke. They watched as the creature glided through the depths, a living testament to their success.

"It's not just water," Avery said, her voice barely a whisper, yet clear enough for them to hear. "It's a future."

They stood together on the dock, four different people brought together by a drying pool of water, now united by a deep, enduring commitment to the land beneath their feet. The spring continued to pulse—a quiet, rhythmic, and constant heartbeat that promised to echo through the community for generations to come.

As the sun dipped below the tree line, painting the water in hues of amber and violet, Avery knew their work was far from over. There would be more meetings, more data to track, and more battles to fight to keep the encroaching world at bay. But as she watched the spring flow, strong and steady against the darkening forest, she felt a profound, unshakable peace.

They hadn't just saved a spring. They had found their place within it. And for the first time in her life, the future didn't feel like a mystery to be feared—it felt like a current, waiting to be followed, clear and bright, winding its way through the heart of the Florida scrub.

The spring continued to rise, its waters lapping gently against the restored banks, a soft, musical sound that signaled the beginning of a new chapter for the Glasswater Spring. They left the dock together, walking back toward the parking lot as the first stars began to pierce the twilight, their footsteps synchronized, their voices blending into the hum of a forest coming back to life. The investigation was finished, but the stewardship had just begun, a quiet, ongoing promise made to the earth, kept one day at a time.

The humidity of a Florida afternoon usually felt like a physical weight, but today, as Avery stood at the edge of the basin, the air felt crisp, almost vibrant. She adjusted her camera lens, the glass cold against her palm. Beside her, the water—once a stagnant, murky amber—had begun to shiver. It was the movement of a heartbeat returning to a dormant chest.

"Look at the eelgrass," she murmured, pointing toward a submerged patch that had looked like dead twine only a week ago.

Micah knelt beside her, his hands caked in the rich, dark muck of the riparian zone. He didn't look up immediately, his focus entirely on the delicate sway of the stems. "It's stabilizing," he said, his voice quiet, stripped of the frantic edge that had defined their late-night planning sessions. "The sediment load from the Enclave run-off has dropped significantly since the bio-retention basins were finished. The spring is finally filtering its own breath again."

A hundred yards away, the rhythmic *thrum* of a hydraulic drill finally cut out, replaced by the shouts of volunteers and the clatter of metal equipment. Declan was standing knee-deep in the secondary drainage channel, his face smeared with grease, holding a tablet up to the sunlight to calibrate the pressure sensor he'd installed near the limestone conduit.

Zara jogged over from the staging area, her notebook tucked into her back pocket. She looked exhausted, her hair pulled back in a chaotic knot, but her eyes held a steady, burning intensity.

"The update from the water board just hit," Zara said, breathless. "They're officially codifying the new land-use regulations. No more aggressive clear-cutting for the next phase of the development. It's done, guys. It's actually in the minutes."

Declan waded out of the water, splashing onto the limestone ledge. He caught the tablet and beamed, a genuine, wide-mouthed grin that looked foreign on his typically stoic face. "The pressure readings are holding. Ever since we cleared that concrete plug near the northern gate, the hydrostatic head has been climbing steadily. We aren't just seeing a spike from the rain—we're seeing a sustained recovery."

"It's not just the water," Avery added, pointing her camera toward a flash of blue near the reeds. A little blue heron landed, stalking the shallows with a precision that hadn't been seen in months. "The hunters are coming back. The ecosystem is recognizing the health of the base layer."

They stood in a small circle, the four of them, looking out over the basin. It had been weeks of back-breaking labor—moving stone, hauling native plantings, and navigating the bureaucratic minefield of the Town Hall—but standing here, it felt like the spring was responding to them personally.

"We have to do a final check of the secondary spillway before the evening squall," Micah said, rising to his feet and wiping his hands on his cargo shorts. "The forecast calls for two inches tonight. If the new basins hold through that, we've proven the system."

"You're acting like a proud father," Zara teased, bumping his shoulder with her own.

"It's the first thing we've built that actually gives back," Micah replied, his tone thoughtful. "Everything else—the roads, the houses—it just takes. This? This is a cycle. We just helped it start turning again."

As they walked toward the northern gate, the sky began to bruise with purple clouds. The Florida storm season was notoriously temperamental, but today, the rain felt like a blessing rather than a threat.

Declan led the way, his stride confident. "The geological data is going to be the most important part of the report I'm writing for the University," he said. "People think the ground is static, that it's just something you build on top of. I think I've got enough evidence now to prove that the karst topography of this region is effectively a living, breathing circulatory system. If you constrict it, the landscape suffocates."

"Make sure you mention the community effort, too," Zara reminded him, clicking a pen. "The data won't mean much if people don't understand that they have the agency to change the outcome. I want this to be a blueprint, not just a case study."

They reached the gate, where the water was already beginning to surge against the newly cleared limestone channels. The sound was a low, melodic roar—the song of a healthy spring. Avery felt it in her chest, the familiar frequency she had missed during the dry spell. She raised her camera and framed a shot: the water cascading over the natural rock, the green of the new vegetation, and in the background, the distant, shimmering heat of the Florida sky.

Avery listened to their banter, a warmth spreading through her chest that had nothing to do with the humid Florida heat. She looked back at the spring one last time. The water was no longer the flat, lifeless mirror it had been a month ago; it was crystalline, revealing the white-sanded floor where the ancient limestone vent pulsed like a steady, rhythmic heart.

They reached the northern spillway just as the first fat, heavy drops of rain began to pelt the canopy. The sound was a symphony—the drumming on the palmetto fronds, the hiss of water against the newly laid limestone riprap, and the gurgle of the drainage system they had painstakingly reconstructed.

"Check the primary intake," Declan shouted over the rising wind, gesturing to the grate they had installed to catch sediment while allowing water to pass.

Avery knelt, shining her waterproof flashlight into the darkness of the trench. The water moving through the channel was clear—a massive improvement over the cloudy, runoff-laden slush that had choked the system for years. She checked the flow gauge attached to the iron grate.

"Flow rate is optimal!" she called out, her voice bright with relief. "The sediment trap is doing its job. No clogging."

"And the nitrate levels?" Micah asked, leaning in close, his face illuminated by her light.

Avery pulled up the latest sensor reading on her phone. "Down by forty percent from last week. The buffer zone is working, Micah. The sedges are doing exactly what you said they would."

The rain intensified, turning the sky a deep, bruised charcoal, but none of them moved. They stood under the shelter of a wide-spreading oak, watching their work hold. The storm, which would have once been a disaster—flooding the roads, washing toxic fertilizers into the spring, and burying the Grotto in silt—was now a source of renewal. It was feeding the aquifer, replenishing the very heart of the spring.

"We did it," Zara said, her voice unusually soft. She reached out and rested a hand on the mossy railing of the boardwalk. "We actually pulled it off."

"We didn't just 'do it,'" Declan corrected, shaking his head as he watched the water swirl into the recharge zone. "We changed the perspective. The town council, the homeowners... they don't see the spring as a property feature anymore. They see it as a lifeline."

Avery felt a profound sense of closure, but more than that, a sense of beginning. She raised her camera and framed the scene: her friends, mud-streaked and exhausted, framed by the lush, rain-drenched green of the spring. She wasn't just observing anymore; she was part of the narrative. She pressed the shutter. The click was tiny, swallowed by the roar of the rain, but the image was captured.

"What now?" Micah asked, looking at the rest of the team. "The restoration phase is a success, but this isn't a 'set it and forget it' situation."

"Monitoring," Zara said, her journalist brain already shifting gears. "We need to maintain the logs. We need to keep the data public. If people stop looking at the spring, they stop caring. We have to keep the conversation going."

"I'll handle the bi-weekly flow reports," Declan agreed. "And I want to expand the drone mapping to the entire northern ridge. If we can map the subsurface cave connections more accurately, we can prevent future developments from encroaching on the main conduits."

Avery looked at the water. "I'll stay with the photography and the wildlife observations. The animals are the best indicators of whether we're actually succeeding or just putting on a good show. As long as the heron stays and the turtles are nesting, we know we're on the right track."

Micah smiled, a slow, genuine expression that reached his eyes. "And I'll keep the native garden project running. I've already got three more neighbors asking how to rip out their sod and plant something that actually helps the water table."

The storm broke as quickly as it had arrived, leaving the park shimmering in the sudden, golden sunlight of a late afternoon Florida sky. Steam rose from the hot asphalt, curling into the cool, damp air of the forest.

Avery walked out onto the wooden dock, the boards still slick with rainwater. She leaned over the railing, looking down into the Grotto. A shadow flickered beneath the surface—a large, dark shape moving against the current. She held her breath.

A manatee, a slow-moving, graceful giant, broke the surface a few yards away, its nose barely brushing the cool, clear water before it submerged again. It was a sign. The spring was open, the connection to the river system was unobstructed, and the habitat was safe once more.

"Hey!" she called out, gesturing to her friends.

They gathered at the railing, their faces lighting up as they saw the visitor. For a long moment, nobody spoke. They watched as the creature glided through the depths, a living testament to their success.

"It's not just water," Avery said, her voice barely a whisper, yet clear enough for them to hear. "It's a future."

They stood together on the dock, four different people brought together by a drying pool of water, now united by a deep, enduring commitment to the land beneath their feet. The spring continued to pulse—a quiet, rhythmic, and constant heartbeat that promised to echo through the community for generations to come.

As the sun dipped below the tree line, painting the water in hues of amber and violet, Avery knew their work was far from over. There would be more meetings, more data to track, and more battles to fight to keep the encroaching world at bay. But as she watched the spring flow, strong and steady against the darkening forest, she felt a profound, unshakable peace.

They hadn't just saved a spring. They had found their place within it. And for the first time in her life, the future didn't feel like a mystery to be feared—it felt like a current, waiting to be followed, clear and bright, winding its way through the heart of the Florida scrub.

The spring continued to rise, its waters lapping gently against the restored banks, a soft, musical sound that signaled the beginning of a new chapter for the Glasswater Spring. They left the dock together, walking back toward the parking lot as the first stars began to pierce the twilight, their footsteps synchronized, their voices blending into the hum of a forest coming back to life. The investigation was finished, but the stewardship had just begun, a quiet, ongoing promise made to the earth, kept one day at a time.

Chapter Twenty-Two

THE RISING SPRING

The rain didn't arrive as a soft mist; it fell as a Florida deluge—a heavy, slate-gray curtain that turned the air into a thick, humid soup.

Avery stood on the edge of the northern spillway, her waterproof jacket slick with moisture. Beside her, Micah adjusted his hood, his eyes tracking the way the runoff from the nearby parking lot was finally behaving. A week ago, this rain would have been a disaster: a surge of brown, debris-heavy sludge rushing directly into the spring, carrying the weight of decades of poorly managed landscape design.

"Watch the lip of the bioswale," Micah murmured, his voice steady against the drumming rain. "The sediment trap is doing exactly what we modeled."

Declan crouched near the ground, his tablet shielded by a custom-built plastic shroud. He didn't look up, his fingers dancing across the screen as he cross-referenced the flow rates. "Velocity is down by forty percent compared to the pre-remediation data. The water isn't scouring the bank; it's sinking."

Zara stood back a few feet, her camera lens wiped dry for the tenth time. She wasn't just capturing the scene for a headline anymore; she was recording the proof of a promise kept to the town. "It's weird," she said, raising her voice over the roar of the downpour. "I spent months trying to find someone to blame for the water levels. Now, looking at this... it feels less like a miracle and more like a logic puzzle we finally finished."

Avery didn't say anything. She was watching the water.

Before the restoration, the water at the spillway would have been murky and agitated, swirling with the chaotic energy of diverted runoff. Today, it moved with a strange, deliberate grace. The new native buffer zones—the grasses and deep-rooted shrubs they had spent weeks planting—were acting as a natural filter, slowing the violent rush of the storm and allowing the water to soak into the limestone base rather than tearing it apart.

"Avery?" Declan's voice was uncharacteristically hushed. "Check the output flow. The pressure gauge at the Grotto vent. It's twitching."

Avery knelt, bringing her face close to the water's edge. She felt the vibration before she saw it. It was a rhythmic, deep-seated pulse, a heartbeat beneath the limestone shelf. For months, the spring had been a gasping thing, a shallow basin of stagnant hope. But as the rain saturated the ground and the groundwater began to recharge the aquifer, the pressure differential was balancing out.

The sluggish, dark water that had been sitting in the basin for days began to push outward. It was being displaced by the clear, cold pulse of the spring.

"It's happening," she whispered.

She stood up, and for a moment, the four of them just watched. The brownish tint of the immediate runoff—the 'storm-water footprint' that had plagued the spring for years—was being outmatched. A plume of crystal-clear, cerulean water pushed from the deep vents, cutting through the surface like a blade. It was the color of the sky after a storm, pure and ancient.

Micah stepped closer, the rain soaking his hair, but he didn't seem to notice. He reached out, his hand hovering over the water, feeling the shift in temperature. "The thermal signature is stabilizing," he said, a slow smile breaking across his face. "The aquifer is finally breathing again."

Zara moved to stand between them, her camera finally lowered. She didn't need to take the shot; she knew exactly how this felt. The skepticism that had once lined the faces of the town council, the frustration of the early days, the sheer exhaustion of the mapping—it all settled into a quiet, profound sense of validation.

"We did it," Zara said, her voice catching. "We actually changed the way the water moves."

"We didn't change it," Declan corrected gently, his eyes fixed on the surging water. "We just gave it back the path it wanted to take all along. We stopped getting in the way."

Avery looked at her friends. They were wet, cold, and exhausted, but the tension that had defined their lives for the last season had evaporated. They were no longer just a group of teenagers clutching maps and shouting into the wind; they were the guardians of the basin. The transition from investigators—people who hunted for the 'why'—to stewards—people who took responsibility for the 'how'—was complete.

"The flow is up another two inches," Declan noted, pointing to the measurement staff they'd installed near the pier.

"It's not just the rain," Avery realized, her voice growing stronger. "It's the lack of friction. The water is actually reaching the spring instead of evaporating on the asphalt or getting trapped in the silt."

She leaned over the railing, looking deep into the center of the spring. A school of small sunfish, which had been hiding in the deeper crevices for weeks, drifted back out into the center of the flow. They moved with the current, confident, steady.

"We have to monitor the runoff all night," Micah said, shifting into the pragmatism of his role. "The peak of the storm is hitting in twenty minutes. If the buffer zones hold during the peak, the project is a success."

"They'll hold," Avery said. She said it with a certainty she hadn't possessed when this started. She knew the way the soil gripped the roots of the muhly grass. She knew how the limestone pores drank the rain. She knew the spring.

Zara leaned her head against Avery's shoulder. "What's next, then? If we're not just 'the kids investigating the drought' anymore, what are we?"

Avery looked back at the spring, watching the clear water continue its slow, steady reclamation of the basin. "I think we're the ones who make sure it never happens again," she replied.

The rain continued, but it no longer felt like a threat. It felt like a renewal. The spring wasn't just surviving; it was beginning to live again, a living, breathing testament to what happened when a community stopped looking away from the ground beneath their feet and started listening to the pulse of the earth.

As the intensity of the storm began to taper, leaving behind the rhythmic dripping of water from the cypress canopy, the four of them stood in silence. The air was cool and smelled of wet earth and wild growth—the scent of a landscape finally at peace. They hadn't just saved a spring; they had found their place within it. The path forward was no longer a mystery; it was a responsibility, and for the first time, it felt like an honor.

The rain had stopped, leaving the air heavy, clean, and smelling of damp limestone and crushed palmetto. For the first time in months, the silence of the spring was not the hollow, eerie stillness of a dying basin, but the rhythmic, melodic pulse of living water.

Avery knelt on the wooden slats of the observation dock, her camera dangling forgotten around her neck. She didn't need to capture this on a sensor; she was capturing it in her blood. She watched as a cluster of eelgrass, which had spent weeks flattened and gray, began to undulate with a sudden, rhythmic grace. It was a slow, tentative dance, but it was alive.

"Look at the turbidity," Declan said, his voice unusually hushed. He held his handheld sensor just an inch above the water's surface. "It's dropping. The runoff from the north ridge isn't carrying the sediment load it used to. The bioswales are actually catching the bulk of it."

Zara leaned over the railing, her tablet open to a live feed of the local community board's public water portal, though her eyes remained locked on the water itself. "The town council just posted the latest flow-rate update," she murmured. "It's still incremental, but it's consistent. No more erratic spikes. For the first time in three decades, the spring isn't struggling to breathe."

Micah was a few feet away, wading calf-deep in the shallow, sandy runoff margin near the bank. He reached down, his fingers brushing against a patch of native river-lettuce that had begun to recolonize the area. He stood up, his face radiant, water dripping from his fingertips.

"The macroinvertebrates are coming back, Avery," Micah said, smiling. "I saw a few caddisfly larvae near the roots. They're sensitive to the nitrates we tracked—the fact that they're settling back in means the water chemistry has fundamentally reset. It's not just a temporary pulse; it's an invitation."

Avery stood up, her joints popping after hours of crouching, and joined them at the railing. The four of them stood in a line, watching the Glasswater Spring. The water wasn't just clear—it was radiant, a crystalline turquoise that seemed to glow from some inner source, reflecting the filtered light of the afternoon sun as it broke through the clouds.

"It's not just the spring," Avery said, her voice steady. "When we started this, I thought we were just fixing a leak. I thought if we plugged the holes and cleared the path, the water would... return. But it feels like something else. The whole park feels different."

"It is different," Declan replied, sliding his sensor into his backpack. "We stopped treating the spring like an object we could manage and started treating it like a partner we had to listen to. That shift? That's what's going to keep it alive. Not just the engineering, but the stewardship."

Zara tapped her screen, pulling up the draft of her follow-up article. It was no longer a frantic report on a crisis; it was a blueprint for continuity. "People are already emailing me," she said, scrolling through the messages. "Not just questions, but offers. The local school's biology department wants to know how we mapped the limestone conduits. The homeowners' association near the north ridge is asking for a list of native plants that don't require heavy fertilization. We aren't the investigators anymore. We're the bridge."

"We're the ones who have to make sure it doesn't happen again," Micah added, his tone somber but hopeful. "The spring is resilient, but it's not invincible. We've learned that the hard way."

Avery looked toward the northern ridge, where the dense canopy of oaks and cypress hummed with the sound of returning birds. They had spent weeks obsessing over data, arguing over maps, and fighting for their right to be heard. But looking at the water now, she realized the most important work hadn't been the protest or the presentation. It had been the quiet moments—the times they sat by the bank, waiting for the water to speak, learning to recognize its moods and its needs.

"I remember thinking the spring was a secret," Avery said quietly. "Something I had to keep to myself because nobody else really saw it. I was wrong. It's a community. Not just the people living in the town, but everything in the basin. We just needed to remind everyone else that they're part of it, too."

Declan walked over to the edge of the dock and dropped a small, weighted line into the deeper section of the basin. The string pulled taut, then drifted gently toward the outflow. "Stable velocity," he reported. "The pressure is holding."

"It's a legacy, isn't it?" Zara mused, leaning her head against the railing. "We're not just saving a spring. We're building a habit. If we keep documenting, keep testing, keep being the ones who notice when the eelgrass turns gray, or the water turns cloudy... we change the baseline. We stop the slide before it starts."

A movement in the shallows caught Avery's eye. A small, dark shape glided through the clear water near the dock pilings. It was slow, lumbering, and utterly content. A juvenile manatee, moving with the current, grazing on the fresh growth near the bank. It nudged a clump of vegetation, unbothered by their presence, an ancient, gentle ghost returning to its home.

Avery pulled her camera up, but she didn't hit the shutter immediately. She watched it for a long, quiet moment, feeling the resonance of the ecosystem—the connection between the rain on the ridge, the flow through the hidden limestone, and the life flourishing in the basin.

"He's back," she whispered.

The others turned, following her gaze. The sight of the manatee brought a collective, breathless silence to the group. It was the ultimate validation. No charts, no historical journals, no legislative agreements could match the simple, profound weight of a creature returning to a place that had almost been lost.

Micah put a hand on Avery's shoulder, his expression a mix of relief and wonder. "It remembers," he said.

"And now, so do we," Zara added, her voice thick with emotion.

They stood there for a long time as the afternoon stretched into a soft, golden twilight. The spring continued to churn—a steady, heart-like beat that promised a future. They weren't just teenagers who had solved a mystery; they were the first line of defense for a world that had taught them how to fight for it.

The twilight settled over the basin, turning the surface of the spring into a mirror of liquid amber and violet. The manatee moved with a slow, deliberate grace, a heavy, submerged shadow that broke the surface once to breathe before drifting back into the depths of the Grotto. It was a silent, living seal of approval on the weeks of labor, the long nights spent pouring over survey maps, and the tense, exhausting hours they had spent advocating in town hall.

"It's not just passing through," Micah said softly, his voice barely rising above the rhythmic lapping of water against the limestone. "It's feeding. That means the flora is healthy enough to support it. The cycle is truly resetting."

Declan packed his gear away, snapping his tablet case shut with a satisfying click. "My sensor logs show the nitrate levels have stabilized within the baseline for a natural spring. The sediment traps held through the deluge, and the flow rate at the spillway is actually slightly higher than our calculated projections. We're not just recovering; we're gaining momentum."

Zara stood with her arms crossed, watching the ripple the manatee left in its wake. She had her recorder tucked away, but she kept a small notebook in her back pocket. It was filled with names—the school teachers who wanted to organize field trips, the neighbors who had volunteered to help with the next round of native planting, and the local shop owners who had agreed to stop using chemical fertilizers on their lawns.

"The article runs in tomorrow's edition," she said, though the tone was casual, as if the news were just a footnote to the scene before them. "It's not an investigative piece anymore. It's a call to action—a guide for how the town can maintain the bio-retention basins. I'm calling it 'The Glasswater Promise.'"

Avery finally lowered her camera. She had taken exactly one photo: a single, crisp frame of the manatee's back breaking the surface, with the clear, turquoise water glowing beneath it like a gemstone. She didn't need a hundred shots. She had the memory of the light, the smell of the rain-drenched earth, and the feeling of the current against her skin.

"I'm going to set up a permanent monitoring station at the pier," Avery said, her voice carrying a newfound weight of authority. "I'll need help with the water clarity samples twice a week, and someone needs to keep an eye on the eelgrass density near the Grotto."

"Count me in," Declan said immediately. "I can run the flow data through the modeling software every Friday. If we see even a ten percent drop, we can cross-reference it with the local rainfall records before it ever becomes a problem."

"And I'll handle the community outreach," Micah added, pushing his wet hair out of his eyes. "We need to make sure the homeowners understand *why* they're seeing these plants, and why they mustn't treat them like weeds. We educate, we facilitate, we protect."

Avery looked at each of them in turn—her team, her friends, the people who had walked into the mud and the dark with her to find the truth of the spring. They were no longer the outsiders. They were the heartbeat of the park.

"The town hall meeting next month," Zara added, her eyes brightening with a sudden thought. "We should present the data together. Not as a report on a crisis, but as a success story for the future. We can show them that when you stop fighting the land and start listening to it, the earth responds."

They walked away from the railing and toward the parking lot, their footsteps crunching on the wet limestone path. The park was quiet, the usual cacophony of the spring's visitors long gone, leaving behind only the sounds of the night—the chirping of cicadas, the distant call of an owl, and the constant, steady rush of the spring water.

As they reached the edge of the lot, Avery turned back one last time. Under the rising moon, the Glasswater Spring looked ancient and vast, a portal into a deeper, cooler world. It was a secret they had been trusted with, a fragile piece of Florida that was now, for the first time in years, truly safe.

"We start the new maintenance cycle on Monday?" Micah asked.

"Monday," Avery confirmed.

"Rain or shine," Declan added with a grin.

They moved toward their bikes and cars, a group of four teenagers who had started the summer as strangers to the complexity of their own backyard and were ending it as the custodians of its future. As they drove away, the headlights cut through the darkness, but for Avery, the real light was behind them—a blue, glowing pulse in the earth that would be waiting for them when they returned.

The work wouldn't be easy. There would be more storms, more seasons of drought, and more moments where the community would have to choose between convenience and the survival of the aquifer. But that was for tomorrow. Tonight, the water was rising. And for the first time in her life, Avery knew that the future wasn't something that just happened to the land; it was something they would shape, one drop, one observation, and one day at a time.

She started her engine, the sound echoing briefly against the trees, and followed the others onto the winding road. Behind her, the spring continued its steady work, the water carving its path through the limestone, clear and cold, feeding the world beneath the surface, exactly as it was meant to do. The drought was a memory; the recovery was a reality. And they were the ones who would see it through.

AUTHOR'S NOTE

Glasswater Spring does not exist. Neither does the Enclave, or the town, or Avery, or anyone she argues with.

The water exists. If you are reading this in Florida, it is underneath you right now.

The Floridan aquifer system is a layer of porous limestone that runs beneath the entire state and reaches into Georgia, Alabama, and South Carolina. It is the primary source of drinking water for roughly ten million people. Florida sits on top of more than a thousand natural springs, the largest concentration on Earth, and every one of them is a window into that limestone. What comes out of a spring is what is down there.

So the hydrology in this book is real. A high-capacity well can measurably draw down a nearby spring. Nitrate from fertilizer and from septic systems does move through karst rock and into spring water, and it does feed the algae that turns a clear vent green. Flow can be measured. Permits are public record. Everything Avery does with a notebook and a stopwatch is something a real person can do.

But I want to be honest with you about the part of this story I invented for convenience, because it is the part that matters most.

Florida's springs are not mostly dying because of one dramatic villain with a pipeline. They are dying because of all of us, slowly, in amounts too small to feel like anything.

The state withdraws something on the order of 3.6 billion gallons of fresh groundwater every day, and roughly sixty percent of that comes out of the Floridan aquifer. About half of it goes to public water supply and a bit more than a quarter to agriculture. All commercial and industrial self-supplied use — which is the category that includes the bottling plants people usually reach for first when they want someone to blame — is a single-digit slice of the total.

Here is the number that changed how I thought about this book. A study of Central Florida homes found that, on average, sixty-four percent of the treated drinking water a household used went onto its lawn. In the summer months that figure rose to eighty-eight percent.

Almost nine gallons in ten. Sprayed on grass. Pumped up from the same limestone that feeds the springs.

We live here. We are a part of the number in this book.

The consequences are already measurable and already being managed for. About two-thirds of Florida has been designated as a water resource caution area, meaning the water available there is not expected to meet the demand projected over the next twenty years. State analysts estimate Florida will need to develop several hundred million additional gallons a day by 2045 just to keep up. And downstream of all of this, in the Indian River Lagoon, roughly three-quarters of the seagrass acreage vanished after 2011 — not from drought, but from algae blooms fed by nutrient pollution, which blocked the sunlight the grass needed to live. The manatees that ate that grass starved.

None of that is a story about a villain. It is a story about arithmetic.

I gave Avery an antagonist anyway, and I want you to know I did it on purpose and with some discomfort. A novel needs pressure and a face to push against. A watershed does not have a face. If this book makes you angry at a corporation and nothing else, I have done only half my job —

and I would rather tell you that plainly here than let you close the book believing the problem is smaller and simpler than it is.

One more admission. Aquifers are slow. A real spring system responds to changes in pumping over years and sometimes decades, not over a single summer. I compressed that timeline so that a seventeen-year-old could see the result of her own work inside the span of a story. If you go looking for a spring that recovers as fast as Glasswater does, you will not find one. What you will find instead is that the recovery takes longer than the damage did, which is the harder and truer version of the same lesson.

And still — this is the part I believe most — springs come back. Silver Glen, Wakulla, Ichetucknee: every one of them has people who show up for it, count things, file comments, plant buffers, and refuse to look away. Florida's springs have been measured and defended by ordinary people for over a century, most of them without a title or a budget.

There are roughly eleven hundred springs in this state. Most of them are not monitored by anyone at all.

That is a problem. It is also, if you want it, a job opening.

Diane Kann

Christmas, Florida

WHAT'S REAL AND WHAT ISN'T

Real

The Floridan aquifer system, its extent, and its role as drinking water for about ten million people.

Karst geology — limestone dissolved by slightly acidic rainwater into the caves, conduits, sinkholes, and vents that make a spring possible.

Localized drawdown. Pumping heavily near a spring can lower the potentiometric surface enough to reduce the spring's discharge.

Nitrate loading from fertilizer, livestock, and septic systems, and the eutrophication that follows.

Consumptive use permits. They are issued by Florida's water management districts and they are public records. Anyone can request one.

Measuring discharge with a flow meter, a float, or a weir — the methods Avery uses are real field methods, simplified.

Citizen water monitoring programs, which supply a meaningful share of the long-term data that exists on Florida's smaller springs.

Invented

Glasswater Spring, the town, the school, the Enclave development, and every character.

The speed of the spring's decline and recovery, compressed for the story as described in the author's note.

The 1922 survey map. These maps do exist but I did not use an actual survey map.

The specific permit numbers, hearing dates, and agency procedures, which are simplified. Real permitting is slower, more technical, and involves more paperwork than any novel should ask you to sit through.

GLOSSARY

Words from the science of this book, in the order a spring would explain them.

aquifer An underground layer of rock or sediment that holds water and allows it to move. Not an underground lake — the water fills pores, cracks, and channels within the rock itself.

Floridan aquifer system The vast limestone aquifer beneath all of Florida and parts of Georgia, Alabama, and South Carolina. One of the most productive aquifers in the world.

karst A landscape formed where slightly acidic water has dissolved soluble rock, usually limestone, creating sinkholes, caves, springs, and underground drainage. Water moves through karst quickly, which means pollution does too.

recharge Water entering an aquifer, usually rainfall soaking down through soil and rock. Paved surfaces reduce recharge by sending rain to storm drains instead.

discharge Water leaving an aquifer. A spring is discharge you can see. Measured in cubic feet per second or in gallons per day.

artesian Groundwater under enough pressure to rise on its own when a path opens. Florida's springs are artesian; the water is pushed up, not pulled.

potentiometric surface A map of the pressure in a confined aquifer — the level to which water would rise in a well drilled into it. When pumping increases, this surface drops.

drawdown The lowering of the water level or pressure in an aquifer caused by pumping. Greatest at the well and decreasing outward, in a shape called a cone of depression.

confining unit A layer of low-permeability rock or clay that slows water from moving between aquifers. Where it is thin or breached, the aquifer below is far more vulnerable.

spring vent The opening where groundwater emerges at the surface.

spring run The stream that carries water from a spring vent to a larger river, lake, or estuary.

first-magnitude spring A spring discharging an average of at least 100 cubic feet per second — roughly 64.6 million gallons a day. Florida has more than thirty, more than any other state.

nitrate A form of nitrogen that dissolves readily in water and moves easily through karst. Sources include fertilizer, livestock waste, septic systems, and treated wastewater.

eutrophication Nutrient enrichment of a water body that triggers excessive algae growth. The algae blocks light and, when it dies and decomposes, consumes the oxygen other organisms need.

turbidity How cloudy water is because of suspended particles. A clear spring going turbid is a signal worth chasing.

dissolved oxygen Oxygen available in water for fish and invertebrates to breathe. Falls when decomposition rises.

saltwater intrusion Seawater moving inland into a freshwater aquifer, often when heavy pumping lowers freshwater pressure enough to let it in. A one-way problem in most cases.

consumptive use permit The legal authorization required to withdraw significant quantities of water in Florida. Issued by a water management district. A public record.

water management district One of five regional agencies that govern water supply, permitting, and flood control in Florida. Springs fall under their jurisdiction.

impaired water A water body that fails to meet state standards for its designated use. Impairment is a legal classification with regulatory consequences, not just a description.

BMAP Basin Management Action Plan — the state's framework for restoring an impaired water body by reducing pollutant loads across everyone who contributes to them.

baseline data Measurements taken before a change occurs, so that the effect of the change can be shown. Without a baseline, a decline is only an opinion.

QUESTIONS FOR DISCUSSION

Grouped by subject so that a science class and an English class can each find their way in. Nothing here requires having read the others.

Science and Evidence

Avery cannot prove anything with a single measurement. Why not? What does she have to do instead, and how long does it take her?

What is a baseline, and why is the absence of one the biggest obstacle Avery faces? Whose job should it have been to collect it?

Trace the path a nitrate molecule takes from a lawn to a spring vent. Which steps in that path could a community realistically interrupt?

The novel treats a public permit as a piece of scientific evidence. Is it one? What can a permit tell you that a water sample cannot, and the reverse?

Karst moves water quickly. Explain how the same property that makes Florida's springs beautiful also makes its drinking water easy to contaminate.

Design the study Avery should have run if she had unlimited time and equipment. What would you measure, where, and how often?

Language Arts and Craft

The spring is described differently in the opening chapter than in the closing one. Find both passages. What specific words changed, and what is the author signaling with them?

Avery narrates. How would this story differ if Zara told it? Choose one scene and rewrite a page in her voice.

Declan and Micah want different things from Avery and from the spring. Are either of them wrong? Argue the position you disagree with.

Water is used as an image throughout — for memory, for pressure, for what cannot be taken back. Track one of these uses across three scenes.

The author's note says that the novel's antagonist is not the real cause of the problem, and that the author knew this while writing. Was that an honest choice or a compromise? Does knowing it change how you read the ending?

Ethics and Civic Action

Sixty-four percent of household water in one Central Florida study went onto lawns. If that is true where you live, who is responsible for the spring — the developer, the utility, the county, or the household?

Avery is seventeen and has no standing, no budget, and no title. Identify every source of leverage she actually uses. Which of them are available to you right now?

Public comment periods, permit files, and agency meetings are open by law and almost entirely unattended. Why do you think that is, and what would change it?

The story ends with a win that does not fix the underlying problem. Is that a satisfying ending? Is it an honest one? Can it be both?

Find out where your own drinking water comes from. Name the aquifer or the reservoir, the utility, and the agency that regulates it. If you cannot find all three, that is itself worth discussing.

WHAT CAN YOU ACTUALLY DO

Not a list of good intentions. These are things a person without money or authority can do this month.

Find out what you are standing on

Look up your water source, your utility, and your water management district. Read the annual water quality report your utility is legally required to publish. It is dull and it is specific, and reading one puts you ahead of most adults.

Turn off the sprinklers

If most of a household's water is going onto grass, this is the single largest lever a family controls. Water less often and more deeply. Replace turf with native plants that do not need irrigation at all. A yard that survives a Florida summer without a sprinkler is not a sacrifice; it is a better yard.

Fertilize less, or not at all

Nitrogen that a lawn does not use goes into the ground. Follow local ordinances on application timing, use slow-release forms if you use any, and never fertilize before heavy rain.

Maintain the septic system

A failing septic system in karst country is a direct line into the aquifer. Inspection and pumping on schedule is unglamorous and it matters more than almost anything else on this page.

Collect data nobody else is collecting

Volunteer water monitoring programs will train you, hand you equipment, and use what you gather. Most of Florida's springs are unmonitored. That is a gap a determined student can genuinely fill.

Show up where the decisions are made

Permit applications carry public comment periods. Water management district governing boards meet in public. County commissions vote on land use. Attendance at these is low enough that one prepared person in the room is noticeable — which is the entire premise of this novel and is not an exaggeration.

Take somebody to a spring

People protect what they have stood in. This is the least measurable item on the list and possibly the most effective.

TO LEARN MORE

Organizations and agencies working on Florida water. Search any of these by name.

Research and Advocacy

Florida Springs Institute

Florida Springs Council

The Florida Springs Institute's Springs Protection reports

Agencies

Florida Department of Environmental Protection — Florida Springs and Aquifer Protection

U.S. Geological Survey — Florida Water Science Center

Your regional water management district: Northwest Florida, Suwannee River, St. Johns River, Southwest Florida, or South Florida

UF/IFAS Extension — your county office, for soil testing, fertilizer guidance, and Florida-Friendly Landscaping

Get in the Water

Florida State Parks — many of the major springs are state parks with free interpretive programs

Florida LAKEWATCH and other volunteer water monitoring programs

CONFIRM: Print names only, no web addresses. Links rot, and a dead link in a printed book is permanent. Verify the Odum entry — it may be the same organization as the Florida Springs Institute rather than a separate listing.

ACKNOWLEDGEMENTS

This book was built on other people's patience.

To the hydrologists, park rangers, extension agents, and volunteer monitors who answer questions from strangers for no reason other than that the questions are about water — thank you. Any accuracy here belongs to you. The errors are entirely mine.

To my family, who have absorbed a great many dinner conversations about nitrate.

To the team — for the covers, the layouts, the tables, the long drives, and for making a small press feel like a real one.

To the librarians and teachers who put books in front of young people who did not know yet that they wanted them. This one is aimed squarely at you.

And to the dogs, who supervised.

About the Author

Diane Kann studied animal science at Cornell University and has spent the fifty years since working with animals, land, and young readers — often at the same time.

She writes from an organic farm in Christmas, Florida, which she runs with her husband and a shifting population of rescue dogs. She has spent decades in animal rescue and training, and she homeschooled nine children, all now grown. Both of those experiences taught her the same thing: that young people will go a very long way into a hard subject if somebody is honest with them about it.

Her writing is what she calls eco-hopepunk — stories grounded in real science and real places, in which the natural world is worth saving and the people in it are capable of saving it. She has written more than eighty books for readers from picture-book age through young adult, including the Signal Ice series, which follows young scientists working at the edge of the world's climate record.

She visits schools throughout Florida.

ALSO BY DIANE KANN

The Signal Ice Series

Young adult. Climate science at the poles.

The Signal Beneath the Ice

Echoes in the Ice

Eco-Adventure for Younger Readers

Middle grade and early chapter books. Real ecosystems, real science, hopeful endings.

The Plastic Tide

The Coral Maze

The Case of the Vanishing Reef

Nami's Journey

Picture Books

For the youngest readers. Several available in bilingual English and Spanish editions.

Rainbow Reef Series and Collection

Not Just! Series and Collection

One Wild Day Series

For a complete list of titles, visit DianeKann.com

www.ingramcontent.com/pod-product-compliance
Ingram Content Group UK Ltd.
Pitfield, Milton Keynes, MK11 3LW, UK
UKHW010011031026
15589UKWH00008B/1592

9 798951 552037